Ten Geographic Ideas That Changed the World

Ten Geographic Ideas That Changed the World

Susan Hanson, editor

Rutgers University Press

New Brunswick, New Jersey

Library of Congress Cataloging-in-Publication Data

Ten geographic ideas that changed the world / Susan Hanson, editor.
    p.   cm.
    Includes bibliographical references and index.
    ISBN 0-8135-2356-7 (cloth : alk. paper). — ISBN 0-8135-2357-5
(pbk. : alk. paper)
    1. Geography—Philosophy.   I. Hanson, Susan, 1943–
G70.T37   1997
910'.01—dc20   96-18123
                                                              CIP

British Cataloging-in-Publication information available

Manufactured in the United States of America

To the generation before us

Ruth Dorrell and Robert M. Easton
Kathleen Berry and Perry O. Hanson, Jr.

all lovers of great ideas

Contents

Preface

Few Americans seem able to forget the "geography" they encountered in the fourth or fifth grade: the challenge of facing a blank outline map of the United States with the charge to name the states, a map with wavy lines on it with the assignment to identify the rivers, or a map of linear lumps with the job to write names on mountain ranges. The tenacity of these memories—correctly associated with geography as we encountered it in elementary school—together with the total absence of any further instruction under the rubric of geography in junior high or high school (and often college as well), places us who are professional geographers in the position of frequently having to define our chosen field. Why would fully grown adults freely choose to spend their days labeling states, rivers, and mountain ranges? What do we do with our time once all the blank maps are filled in? The same memories seem also to require us frequently to defend our chosen field: Why does the world need geography? What have geographers really done for the world (beyond labeling all those features on maps)?

This book grew out of the plenary session I organized for the annual meetings of the Association of American Geographers (AAG) when I was president of the association in 1991. My motivation for setting up such a session was not to define geography but to stimulate people in our field to think about what geographers have contributed to the world and how geographic thinking has changed, and might change, the world. I wanted us, as a discipline, to reflect on intrinsically geographic ideas that had had an impact outside the academy. The session consisted of three presentations, prototype versions of the chapters included here by John R. Mather, Robert Kates, and Patricia Gober, and an invitation to the audience to propose additional great geographic ideas.

This collection of ten geographic ideas that changed the world gradually emerged from that beginning. In the process, the goal of the book broadened beyond that of the AAG session to encompass introducing nongeographers to the world of geography. Some of the contributors contacted me with their proposed ideas, and in other cases I solicited contributions on particular topics. In all cases, the enthusiasm with which this group of prominent geographers has embraced this project has been remarkable and gratifying.

These ten are not the only geographic ideas that have changed the world; they might not even be the ten most important geographic ideas ever to have emerged. Each does, however, have its origins in one or more core geographic concept, and each has been around long enough to have had an impact on the world beyond our discipline and on the world outside the walls of universities. Our intention is not, in these few brief chapters, to survey the field of contemporary geography; rather, our hope is that this volume will give newcomers to the field some glimpses of geographic ideas that literally changed the world and our thinking about it. The emphasis throughout is on the ideas and their significance both inside and outside of geography, rather than on the ideas' originators. Some readers might argue with good reason that other great geographic ideas should have been included; in this vein the idea of diffusion or that of the mental map might be proposed. For a variety of reasons not every "candidate idea" made its way into this collection. In addition, in a few years' time perhaps an idea now just emerging will have had a sufficiently broad impact to qualify as a geographic idea that has changed the world. I welcome such suggestions and invite proposals for possible future editions of this volume.

As we worked on this book, we held in mind an audience of nongeographers—advanced high school students, students at the beginning of their college years, or those who managed to escape from college without any exposure to geography—anyone who has not yet been introduced to the geographer's ways of looking at and understanding the world. Our aim has been to craft a "jargon-lite" book that is accessible to people who have not had any formal advanced education in geography but who want to understand what it is that post-fourth-grade geographers do and how geographic thinking contributes to the world.

To this end, the project became something of a cottage industry, as I enlisted the help of family members to serve as representatives of the population of nongeographers we hope to reach. My daughter, Kristin Hanson, and father-in-law Perry O. Hanson, Jr., in particular, gave us the benefit of their lay perspective and provided valuable suggestions as to how we might convey complex ideas with greater clarity. Their participation not only improved the final draft significantly; it made the whole process of getting there a lot more fun. Many thanks to both of you and also to Bill Meyer, of the Marsh Institute at Clark University, who read several chapters; we benefited from his wisdom and his superb editorial skills. As always, I greatly appreciate the advanced word processing skills of Diane LePage.

Without Karen Reeds, our editor at Rutgers University Press, we would have no *Ten Geographic Ideas That Changed the World*. Karen saw the potential for a book in that initial AAG session and has pursued the idea with vision, lucid insights, persistence, and, above all, patience. Her keen intelligence and broad knowledge of geography have guided the project from its inception, and all of us who have contributed to the book have benefited from her leadership. Although her name does not appear on any chapter, each one bears the stamp of Karen's thought.

Susan Hanson
Worcester, Massachusetts

Ten Geographic Ideas That
Changed the World

Introduction
Ten Geographic Ideas That Changed the World

Susan Hanson

Talk to geographers about almost anything—the Civil War, environmental degradation, traffic congestion, crime, job loss—and they will secretly, or not so secretly, yearn for a map. Geographers want to know not just where things occur, and what the spatial pattern looks like, but why they happen where they do. The answer often lies in a map.

Take the Civil War for example. A geographer will want to know why a battle—say Gettysburg—was fought where it was. But what do we mean by "where"? The latitude and longitude of the battlefield tells us where, in one sense, but by itself latitude-longitude location is a distinctly unsatisfying answer to where. More revealing will be a map of the eastern United States showing topography, transportation routes, and settlement patterns. Such a map will reveal that the battle of Gettysburg was fought where it was because General Robert E. Lee, planning to outflank Union troops and attack Washington, D.C. (the Union capital), from the north, had moved his Confederate troops northeast through the protected transportation route of the Shenandoah Valley. By the time he reached southeastern Pennsylvania, the Union army was in pursuit, and when both sides ordered their troops to assemble at a central location, the Battle of Gettysburg ensued. Why? Because all roads in the area led to Gettysburg. The transportation geography of the region meant that both armies converged on Gettysburg, where they encountered each other unintentionally in a "meeting engagement" that became one of the Civil War's bloodiest battles (see Figure 1).[1]

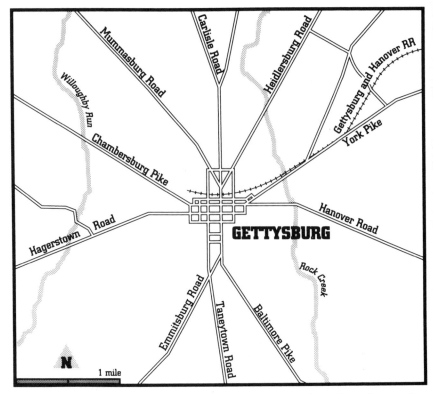

Figure 1. The gathering of regional highways in Gettysburg brought together the Confederate and Union troops in a "meeting engagement" at the end of June 1863. Source: Adapted from Winters (1992:166).

This example illustrates how a geographic turn of mind, in posing different questions (what some have referred to as "the why of where"), leads us to consider different explanations. Something about the geographic turn of mind wants to see the big picture, is not content with unrelated fragments, and wants to grasp how the pieces fit together in place. The map can provide this synthesizing framework—a touchstone of geography. Show any geographer a map, and she is immediately immersed in relationships and connections—connections between people and the environment, connections between and among places, connections between people and places. These are core concerns of geography.

The word "geography," descended from the Greek, literally means writing about, or describing, the earth. This book is about how the study of the earth has sparked ideas that have literally changed the earth that we study. The changes are always wrought first, and most powerfully, in the intellectual realm, altering how we see and think about the world or some aspect of it. An example is the shift in world view initiated by George Perkins Marsh in the late nineteenth century, from seeing human transformation of the environment as signaling progress to seeing such transformation as signaling potential disaster (a change described in Chapter 6 of this volume).

An idea that significantly alters the intellectual terrain is likely to be translated into changes in the material landscape as well. For example, the idea that human transformation of the environment does not necessarily enhance human quality of life has underwritten a variety of environmental protection laws, such as those to protect wetlands. The design of the settlement system on the Dutch polders, the land the Dutch have claimed from the North Sea, is an execution of the geographic principles embodied in central place theory (described in Chapter 8), and the absence of dwellings in the floodplains of certain rivers in the United States is a visible expression of government policy stemming from the idea of human adjustment to environmental hazards (described in Chapter 4). When geographic ideas change the world in our heads, the impact can sometimes be read in the landscape.

The linkages between visible changes "on the ground" and a shift in how we see the world are sometimes subtle, however, and not always easy to draw. The idea of the map, described in the next chapter and perhaps the most influential geographic idea of all time, has probably changed the world most by changing how we define ourselves.

The ten geographic ideas described here are not the only ones that have changed the world; these ten are not, one might argue, of equal breadth and scope, nor are they necessarily the ten that have had the greatest impact. They do, however, represent significant areas of geographic inquiry and illustrate the ways in which geographic ideas have contributed—and continue to contribute—to shaping the world geographers study. In this book, the chapters describing the ten ideas are organized into three sets of three plus a coda. Within each chapter, the author first outlines the idea and its origins before describing its intellectual development and the nature of the idea's impact inside and outside of geography. The first set of ideas has to do with how we frame the world, how we represent it in maps and other graphic images. The second, addressing connections between people and the environment, focuses on how we see the world as human habitat, one that we humans have transformed and that has transformed us. The third set of ideas, highlighting the connections between and among places, deals with how we have come to see the world as a linked mosaic. The final chapter, on sense of place, tackles connections between people and places.

Frames of the World

How we portray the world not only reflects what we are able to see; it also enables us to see and understand the world in a new light—and perhaps thereby to change it. This theme pervades the three chapters in this first section; each presents an idea having to do with how geographers frame the world—the idea of the map, the idea of a large-scale weather map, and the idea of a geographic information system (GIS).

Perhaps the most quintessentially geographic form of representation, the map allows us to simplify the complexities of the world and to impose organization on it. From its earliest origins, perhaps as a pictograph in the dirt (this stick is the river, the pebble is that tree, and these stones are the mountains over there), the map has remained a powerful metaphor, reflecting and in turn shaping how we see the world. Before the idea of the map, people had to rely on the linear medium of language to convey spatial information. In the symbolic representation of a map, we can impose the ideas of order and hierarchy over two or three dimensions. Accessible across barriers of language and culture, maps often appear unbiased, but they are subtly (and sometimes not subtly) infused with ideology. In the first chapter, Anne Godlewska explores the idea of the map and some of the key developments and elaborations it has undergone, including solutions to the problem of representing a spherical earth on a flat surface, the emergence of the idea of uniform scale, and the evolution of the symbolic language of maps. Can we imagine a world without maps?

Perhaps the map that is hardest to imagine living without today is the weather map. The idea of the weather map expands the fundamental idea of the map in several directions. First, the weather map, with its central concern for air masses, rises above the surface of the earth. Second, representing weather data for a particular, brief period of time over very large areas and using data that are extremely time-sensitive (today's and tomorrow's likely weather are of far greater interest than is yesterday's), the weather map requires the timely assemblage of weather data for a particular point in time from a large number of widely dispersed locations. Third, these locations must somehow be linked together to create the weather map. Realizing the idea of the large-scale weather map required not only the ability to measure temperature and barometric pressure but also the coordination of a national (and then international) network of observers.

In Chapter 2, Mark Monmonier describes the idea of the weather map and outlines its development, explaining how rudimentary maps of temperature, pressure, and precipitation have evolved into the dynamic images that now enliven television weather forecasts. As Anne Godlewska noted in the case of the map, depiction both reflects and shapes our understanding of the world: the synchronous weather map enabled understanding of the effect of atmospheric pressure on wind direction and speed. Monmonier notes how the idea of the weather map has further changed the world by altering our sense of connectedness with distant places and by feeding our sense of control, if not over the weather itself, at least over its worst effects.

The geographer's newest way of framing the world is through a geographic information system (GIS), which enables the collection, storage, and analysis of spatial information in digital form. Not just a new, more high-technology map, the idea of GIS stretches the original map idea even further than does the weather map. Geographic information systems represent a fundamentally new way of thinking about spatial information because that information comes to us, and is entered into a GIS, in highly disaggregated units (a single tree, a house, a road segment, a 10-meter by 10-meter patch of the earth's surface captured in a satellite image), which can be combined and manipulated electronically in an almost endless variety of ways.

As Michael Goodchild points out in the chapter on GIS (Chapter 3), how the world is represented on a map (what features are selected, how they are categorized, and how the earth's curvature is distorted when it is forced onto a flat surface) depends on the decisions of the mapmaker *and* on the available technology. With pen and paper technology, for example, it is hard to depict continuous variations, such as those in the species composition of forests or in the socioeconomic characteristics of neighborhoods. As a result, maps tend to employ (and therefore to reify) crisp boundaries, to the neglect of fuzziness and uncertainty. Can we imagine a world without boundaries as we know them, a world in which the hard lines with which we are so familiar are replaced with soft, fuzzy zones that can change rapidly as the world changes? Can we imagine a census without census tracts, forest management without stands, a world without nation states?

GIS technology forces us to think more carefully than we have in the past about representing continuous change and fuzzy boundaries across space and time rather than sharp distinctions and to convey uncertainty more systematically than we have been accustomed to doing. With examples like these, Goodchild describes how GIS pushes us away from the map metaphor that has become an integral part of our lives and demands that we think anew about many taken-for-granted ways of looking at the world. GIS is an idea that is ever more forcefully now changing the world. These geographic ways of framing the world, of representing it, permeate the geographer's approach to understanding the world as human home and the world as a linked mosaic.

The World as Human Home

The relationship between people and the environment is an enduring theme in geography. The environment has played a powerful role in shaping human life around the globe; at the same time, we humans have been transforming the earth for as long as we have lived here. Each of these themes—human adjustment and human transformation—is a geographic idea that has shaped, and continues to shape, people's relationships to the environment.

The idea of human adjustment—that people can and do mitigate environmental hazards by adjusting their practices and behavior—had its origins in Gilbert White's insights into the ways people have adjusted to living in floodplains. By altering their behavior (for example, by providing emergency warning and evacuation, by building levees and dams), people can avoid adverse environmental impacts. The environment poses many challenges in addition to floods—drought, hurricanes, earthquakes, climate change—but humans are adaptable and capable of devising a broad range of accommodations. This is the message of human adjustment, which Robert Kates describes in Chapter 4. It is the message that human coexistence with nature is often more effective than human dominance over it. Successful adaptation requires that people perceive the environment as hazardous, that they identify the particular populations and locations at risk, and that they determine the adjustments that are possible for those at risk to make. The ability of households or larger groups to adapt in the face of environmental threat depends on their access to power and resources.

The idea of human adjustment has had a powerful impact on public policy, in its insistence, for example, that policy should recognize all possible adjustments and that each of them has far-ranging social costs and benefits. As Kates points out, many of the environmental hazards challenging human ingenuity and adaptability are human-induced; sometimes—as in the case of levees—they are themselves part of an adjustment strategy. Should we take steps to prevent an environmental change, such as global warming, by regulating fossil fuel consumption, or should we rely on people's ability to adapt to whatever changes we may help to create? This question defines the terms of debate between preservationists and adaptationists.

A geographic idea that has proved to be an especially effective tool for human adjustment is that of the water budget. In order to understand the hydrologic cycle and also to classify climates, we needed to be able to assess the relative moistness of different places. This assessment depends upon comparing the available supply of moisture (from precipitation) with the demand for water (through evapotranspiration) in different places. As John R. Mather describes in Chapter 5, water budget climatology quantifies how precipitation is used for evapotranspiration, runoff and streamflow, and the recharge of soil moisture.

The idea of water budget climatology is based in Warren Thornthwaite's insight that *actual* evapotranspiration from an area covered in vegetation is different from *potential* evapotranspiration, which is the amount of moisture that would be lost through evapotranspiration if the vegetation were always well supplied with water. What Thornthwaite saw, therefore, was that the demand for moisture is at times independent of precipitation and depends not just on how much water evaporates, but on how much *could* evaporate. Thornthwaite discovered, moreover, that potential evapotranspiration is independent of vegetation or soil type and depends only on the amount of solar energy available; it is, therefore, a true climatic factor.

Mather describes how this discovery has made possible the calculation of quantitative values for water deficit and surplus for any location in the world. Humans have interfered with the hydrologic cycle through, for example, the draining of wetlands for agriculture and the damming of rivers for reservoirs; because it is based in an understanding of the relationship between climate change and water resources, the climate water budget allows us to assess the impact of human activity on water resources. Successful crop, water, and soil management now depend on the idea of water budget climatology.

The fact that we humans can adapt in many ways to the environment shows that we have never been passive victims of an environmental imperative. Human transformation of the world—evident in cities, air pollution, deforestation, and desertification—has accompanied human habitation of the world. As William Meyer and B. L. Turner, II, point out in Chapter 6, human action, although certainly not the only force altering the earth, is now the principal one. *Ideas* about human transformation have driven people's interactions with the environment and have been revised in light of the outcomes. People can see an earth transformed as either an earth damaged or an earth improved.

Meyer and Turner describe the role of George Perkins Marsh in turning people away from the idea that human transformation of the earth—human conquest of nature—need be equated with progress. They note that we have gone from seeing human modification of the earth as a sign of progress to seeing it as a sign of decline. This shift happened in large part because we have also come to appreciate that, as Marsh emphasized, human impacts are often unanticipated or are felt in times and places far distant from the site of the original modification. We now know that such unforeseen and uncontrolled consequences are just as significant as those that were undertaken by design; in fact, most contemporary environmental changes were not planned or intended. More controversial than the geographic idea of human adaptation, the idea of human transformation now permeates environmental politics as the idea of human adaptation governs environmental policy. Meyer and Turner do not advocate the cessation of all human transformation of the earth; they do point to the need to be able better to foresee the consequences of such actions as a precondition to envisioning viable solutions. The myriad ways that people have transformed the environment and adapted to it have created countless varieties of the world as human home; a web of interdependence knits these worlds together.

The World as Linked Mosaic

Whereas the ideas in the previous section focus on human interactions with the "natural environment," this group of geographic ideas has to do with human interactions across space. At the core of all three ideas in this section (the idea of the functional region, central place theory, and the idea of megalopolis) are the nature of and basis for interactions among disparate and often distant places.

Just as historians have invented ages or eras that group events in time, geographers long ago invented regions, which group events in space. Both eras and regions are a means of imposing order, of seeking for pattern, of creating categories for understanding. We are all familiar with the idea of a region that unites places on the basis of some shared trait, some common similarity. An example is the Corn Belt in the United States, which is comprised of, say, contiguous counties that share certain characteristics associated with corn-based agriculture. Quite different, however, is the idea of the functional region, which defines the basis of membership not in some similarity but in the linkages that people and institutions have developed between and among places (the region *functions* as a unit because of these connections).

As Edward Taaffe notes in Chapter 7, the linkages at the core of the idea of the functional region reflect and create specialization and interdependence. A key insight of this idea is that differences rather than similarities between places can lead to stronger connections, as when one place produces mangoes and another, lentils. As the strength of a linkage increases, place-based specialization, and with it interdependence, are spurred. Taaffe points to the tension between regions based on similarity, such as a high-income suburban enclave, and regions based on difference and the realities of interdependence, such as the whole metropolitan area. He reminds us that only a fool would think that a region defined by homogeneity can be separated from the larger web of functional connections; yet such a delusion seems to be driving separatist inclinations in the Balkans and elsewhere.

Inherent in this idea of the functional region is the notion of hierarchy—that certain places (usually larger ones) will be dominant over others (usually smaller ones). New York and Los Angeles, for example, are important, dominant nodes in the commercial and financial networks that knit the United States, and in some sense even the globe, into a functional region. At smaller scales, Indianapolis and Worcester serve as the centers of much smaller functional regions. The idea of the functional region prompts questions about the advisability of having economic and political power concentrated in ever fewer and ever larger centers.

Taaffe describes how, by focusing on connections, the idea of the functional region has changed the world by shifting attention to differences among places as a basis of interdependence and by defining closeness in terms of linkage rather than geographic proximity and similarity. This idea has also made us realize that the effects of distance are not absolute but relative—relative to the ease with which distance can be overcome. In addition, the idea of the functional region shifted attention within geography away from the uniqueness of areas and toward more general spatial processes, which are applicable across different places.

A direct descendant of the idea of the functional region is central place theory, which has to do with the location, size, economic characteristics, and spacing of market centers. Central place theory focuses on linkages between consumers, clients, or patients on the one hand and markets, clinics, or hospitals on the other. The size of a facility, the size of its market area, the ease of traversing distance, and the distance between facilities of similar size are all related. In Chapter 8, Elizabeth Burns describes the origins and development of the idea that, in the 1930s, German geographers Walter Christaller and August Lösch formalized in central place theory.

Burns notes that this idea has had many extremely practical consequences, primarily in the design and development of cities and towns. When the Dutch were faced with designing an entire settlement system on the brand new polders, they drew upon the idea of central place theory. The idea guides problem solving in established settlements as well. Where should the new fire station be located? What hospitals should be consolidated or closed? How large a shopping center should be built at this location? In solving these problems, planners mobilize the ideas behind central place theory.

Megalopolis, recognized and named more than thirty years ago by French geographer Jean Gottmann, is one form of functional region. As Patricia Gober outlines in Chapter 9, Gottmann saw a new form of urban growth in the interchange of ideas, people, money, and goods that linked the cities in the American northeast into a single functional region. In the size of this gargantuan urban settlement Gottmann perceived the blurring of the boundary between rural and urban. Not only does the scale of megalopolis exceed that of the mere metropolis, but this new form of urbanization emerges from a different economic base—an information-based, service economy rather than a goods-based manufacturing economy.

The essence of this geographic idea lies not only in calling attention to, defining, and naming this new phenomenon, but in seeing the connections between the new urban form and a new basis for the urbanization process. The "old" view of urbanization saw it as closely tied to industrialization, saw manufacturing as *the* engine of urban growth, and ultimately saw wealth being rooted in manufacturing and the natural resources that feed it. Gottmann's insight about urban growth was that services, not just the production of goods, could be the engine of growth and the source of wealth. Human resources, and the linkages and interdependencies among people and places, are central. Gottmann saw the original megalopolis, the U.S. northeast, in particular, as a hinge linking the national with the international economies. The idea of megalopolis has redefined the terms of urban analysis.

Coda: Sense of Place

The linked mosaic, the earth as human home, and geographic frames of the world come together in what geographers call sense of place. By grounding us in specific contexts, in specific geographic homes, a sense of place roots us to the world. Yet a well-developed sense of place—an identity forged from the shared symbols, experiences, and meanings that come from living in one place—is full of ironies, for not only can it connect us to one locality; it can also provide us with the basis for connecting to the wider world. In this sense, it is as if we have to know where we are before we can connect with the rest of the world. In another twist of irony, a sense of place, in grounding us in one part of the world, can divide us from other people and places, as when a heightened sense of local identity feeds hostility toward those who are perceived as outsiders.

These ironies spring from what sense of place is all about. In the final chapter Edward Relph describes how a sense of place is in part "an innate facility" but can also be a "learned skill for critical environmental awareness," a skill that hones one's ability to see the general in the particular. Every place is a combination of the general and the specific; Relph gives, as example, Ruskin's description of a nineteenth-century industrial city in Britain, but we could as well focus on the high-technology districts of Massachusetts and California to show how the general processes of technological innovation are played out differently in different places (see, for example, Saxenian 1994). People with a well-developed sense of place appreciate the shared properties of places and the general processes that produce the specific conjunction of properties that make a place distinctive. In addition, people with a well-developed sense of place appreciate the role that linkages (or the absence of linkages) between and among places play in creating the collage of places that join the general and the specific.

Relph points out that geographic learning fosters this healthy sense of place in at least two ways. First, geographers take their students into the field and teach them how to read a place. Through careful observation, students learn to see connections among the elements of a landscape—working-class and middle-class housing, factories, toxic waste dumps, sewage and water treatment facilities, battered women's shelters, offices, and market gardens. Second, geography educates students about the characteristics of many different places. By helping people to see what is common among places and to understand why places differ, geography enables people to make sense of the diversity among places.

Geography, then, nurtures sense of place as a "learned skill for critical environmental awareness"; in this way, sense of place is not so much, like the other ideas in this book, a geographic idea that has changed the world as it is a skill that undergirds a sensitivity to and understanding of how the world is changing. How is it that, in an era of global connections, local place-based identities are flourishing and all-too-frequently fueling the exclusion and rejection of those who do not "belong" in that place? How does a sense of place illuminate the decisions of those powerful few who spend large portions of their lives circling the globe in airplanes and staying in hotels remote from home?

In building an understanding of how sense of place informs change, Relph highlights three eras—the premodern, the modern, and the postmodern. In the premodern era, when travel was difficult and costly and connections among places were weaker than they are today, a sense of place was grounded in a single, circumscribed locality. Diversity among places was great, but knowledge of distant, different places was little. The modern era, Relph argues, was typified in Bauhaus architecture, the universal, unadorned "box" that came to dominate urban skylines around the globe. Place became irrelevant. The globalization of economic activity, and particularly the stronger linkages among places etched by such activity, erased distinctions among places. Homogeneity reigned as the universal (everywhere or anywhere) asserted its power over the particular. In the postmodern era, which Relph sees represented in Las Vegas, new places are created as collages of place fragments that have been uprooted from their diverse home locations and reassembled in pastiches that seem designed to surprise. Weaving your way through the postmodern landscape and making place sense of it requires, indeed, an advanced understanding of sense of place. Relph concludes that such an understanding should enable us to "live comfortably with difference and to appreciate both what is shared and what is distinctive in different cultures and places."

The themes that run through these ten great geographic ideas have to do with relationships, with connections, with linkages, with interdependencies: connections in place, between the human and the physical worlds, and across places. These connections operate at all geographic scales, from the extremely local (within households and neighborhoods) to the global. They can, ironically, *erase* difference and homogenize places, as in Relph's example of modernist architecture, or they can *create* difference, as in Taaffe's example of ever-stronger linkages feeding ever-greater specialization and complementarity between places. The connections established within a place can nurture an understanding of shared experiences and a sense of belonging, but those same connections can foster ignorance of and hostility toward other places and other cultures. Place can both connect and divide.

Yet the divisiveness of place, based as it is in insularity, exists only in people's denial of the reality of interdependence: The connections are always there; the boundaries are always porous. In her brilliant book *Dakota: A Spiritual Geography*, poet and essayist Kathleen Norris (1993) reminds us that we ignore linkages at our peril, for such ignorance blinds us to the process of change and leaves us powerless to cope with change. Reflecting on the decline of the agricultural economy of the Dakotas, Norris wryly comments on the complacent and self-destructive insularity of Plains people: "Paradise wasn't self-sufficient after all, and the attitude and the belief that it ever was is part of the reason it's gone" (p. 47). She describes the people of her small South Dakota town as strongly resistant to change and traces that resistance to their refusal to connect with the outside world "except through the distorting lens of television" (p. 50), to their hostility toward and distrust of outsiders, and to their eagerness to idealize their isolation. In eschewing connections, she observes, the people of her community "foresak[e] the ability to change" and in so doing "they diminish their capacity for hope" (p. 64). The ten geographic ideas described in this volume have changed the world by emphasizing our interdependence on it and in it; in so doing they have enhanced our intellectual flexibility and our ability to change.

Note

1. I am indebted to Professor Harold "Duke" Winters for this example, gleaned on one of his legendary field trips to Civil War maneuver sites and battlefields.

References

Norris, K. 1993. *Dakota: A Spiritual Geography.* New York: Ticknor and Fields.

Saxenian, A. 1994. *Regional Advantage: Culture and Competition in Silicon Valley and Route 128.* Cambridge: Harvard University Press.

Winters, H. 1992. Geography and the Civil War—The Eastern Theater and Gettysburg. In *The Capital Region: Day Trips in Maryland, Virginia, Pennsylvania, and Washington, D.C.,* ed. Anthony R. DeSouza, 137–177. New Brunswick, N.J.: Rutgers University Press.

PART 1

FRAMES OF THE
WORLD

1

The Idea of the Map

Anne Godlewska

Of all the ideas explored and developed by geographers over the last several thousand years, the map has become the most central to Western civilization. Once the privileged instrument of monarchs and prelates, it is now the everyday tool of magazine readers, weather watchers, mall users, museum visitors, and the dwellers of labyrinthine modern office buildings. Not so long ago the exclusive responsibility of geographers, maps are now made by journalists, scientific specialists, graphic artists, and, indeed, anyone with a computer mapping package. As astronomers attempt to chart the universe and geneticists strive to map genes, maps are, in addition, an almost all-pervasive popular image representing power and control; they also capture a sense of "the joy of being there" in advertising, on placemats and on postcards. In the last two hundred years, maps have been both demoted and elevated to the everyday. As such, they are one of geography's greatest success stories.

All maps are not born equal. The placemat map and the topographic map play radically different social roles, depending not only on their intrinsic nature but also on the context of their production and use. Indeed, when read carefully and with critical attention to their contexts, maps can tell us a great deal about the societies within which they reside. But just as most consumers of electrical power know nothing about the principles of electricity, most modern map users (and many modern mapmakers) know nothing of the map's history and understand little of the subtlety with which the map speaks. This, despite

Figure 1.1. Although this painting uses both cartographic and pictorial elements, its depiction is impressionistic and evocative rather than informational. It does not have a consistent symbol system, there is no linguistic code, the location of the symbology does not seem critical, and the

the fact that the map's history is in many ways our history. Central to many of our most astonishing discoveries, from continents to the earth's history, the map is also a testament to our overweening desire to control and dominate others, nature, and even our past.

The Map—What a Great Idea!

The map is also one of the oldest and perhaps the most powerful and constant of geographic ideas. Maps express facts or concepts that derive a large part of their significance from their spatial relationships. Although they may be as beautiful as any work of art, we distinguish maps from art in the way we look at them. In a work of art each element contributes to the total effect, and it is the totality, its meaning to us, and our ability to find resonance with it that gives it power and a voice. It is possible to look at a map in this way. We have all seen maps of the world, maps of estates, county maps, and maps of the home

tableau has no geometric frame. Most readers will judge it to be more art than cartography. David Hockney, *Mulholland Drive: The Road to the Studio*, 1980, acrylic on canvas (86 × 243 inches). © David Hockney.

town framed and prominently displayed. Indeed, many old maps seem more pictorial to us then representational. It is also possible, as David Hockney has demonstrated (Figure 1.1), to create art that mimics some of the symbolism and even the structure of a map.

For most map readers, certainly for most readers of contemporary maps, the map's message does not lie in its overall effect but in the locational information it carries, conveyed by its complex array of symbols and the relationship between those symbols in their relative location in the geometric frame of the map. Most map readers never look for, and never hear, a voice in the map. In a sense, then, a map is also born of an agreement between the mapmaker and the map user. The map user understands the mapmaker to have represented a certain territory with fidelity and according to accepted principles. This understanding shapes and directs the map user's gaze.

The distinction, however, between art and cartography is perhaps more real today than at any time in the prehistoric past, when maps were probably used not only analytically and illustratively but to conjure. Consequently, in the absence of detailed knowledge of prehistoric societies, and with no understanding of either the map reader or the mapmaker, it is often impossible to identify and interpret prehistoric maps with any certainty. Still, historians have persisted in their attempts to do so precisely because it is hard to imagine a time or a culture in which maps were not present in one form or another. Nevertheless, although we may consider cartography distinct from art, mapmaking is a creative activity demanding artistic skill and judgment, often in addition to significant mathematical and analytical ability.

Because maps have had, and continue to have, many forms, generalization is problematic. Perhaps the most fundamental distinction is between the conceptual map, or mental map, and maps that have a physical manifestation. A mental map is the image of space with which an individual functions on a day-to-day basis. In contrast, the drawn or printed map is part of a dynamic of communication and exchange. Shaped by the individual's own highly personal experience, needs, and emotions, the mental map will differ significantly from any modern printed map. How we conceptualize space (and, therefore, the mental map) is only dimly understood and still barely explored, in spite of the work of Kevin Lynch (1960) and Peter Gould (1974), the subsequent work of psychologists in space perception and mental mapping (McGuinness 1992), the recent work of artificial intelligence researchers, and the thoughtful writings of Christian (Jacob 1992). Yet there is every evidence that all functioning adults and mobile children compose mental maps at a dizzying rate every day and probably every night of their lives (Castner 1990; Wood 1992).

Maps with a physical manifestation also have an enormous variety of shapes and functions. They can be three-dimensional models and still be maps. They can be statements on, or attempts to fathom, the configuration and nature of the cosmos. They can be way-finding tools expressed in any medium: paper, sand, bark, papyrus, animal skin, air (drawn with a finger perhaps), snow, oil on cement—any medium at all. They may describe the world as perceived or the world as imagined or dreamed. Most often drawn for the living, such maps have also been composed to assist the dead in finding their way in the afterlife. They may be encyclopedic and, like the topographic map, attempt to be all things to all people. Maps may simultaneously—and more or less obviously and deliberately—seduce, confuse, and obfuscate. Indeed, on occasion, as when we find them in advertising (broadly defined), maps may have a primarily symbolic or ideological function.

They may be invaluable and utterly necessary tools in the planning of complex parts of communal life such as air transport routes, and telecommunication systems or in situations of war or the hunt where the concerted exertion of force is of crucial importance. Maps have been used to investigate and conceptualize the path of disease, the distribution of plant and animal life, the structure of rock, the evolution of the planet and the solar system, and the structure of the brain and genes (Hall 1992), and they have been essential in the exploration of any number of other intellectual problems. Finally, because they are frequently used as a basis for ruling, directing, controlling, and limiting the movements and actions of others, maps are a locus of considerable civic and religious power (Godlewska 1994, 1995). It is worth remembering that the meanings of these maps are clear to us only insofar as we know something of the context of their social production; the prime function of the historian of cartography is to elucidate these contexts (Harley 1989; Wood 1992).

The Origins of the Map

Of all the geographic ideas discussed in this book, the map is the one the origin of which is most difficult to isolate. Based on what little we know of human cognitive development, sparse archaeological evidence, and the little that historical linguistics and semiotics can tell us, we think that mapping is older than writing and mathematics but younger than music and dance, and that it has links to the development of gesture-based communication (McNeill 1992; Raffler-Engel et al. 1991; Wind et al. 1989). Some scholars working on the origins of language speculate that mapping predates *Homo sapiens* and the formation of the modern human brain (Hewes 1977; Kendon 1975). By analogy from nonliterate peoples and in particular from the sign language and sand paintings of the Walbiri in North West Australia and the sign language of the North American Plains Indians, others argue for an association between gesture, icon, pictograph, map, and the very earliest elements of writing (Meggitt 1954; Munn 1966; Kendon 1981). Still others, using the ideas of Piaget, have explored the possibility of understanding early human cognition by analyzing the spatial cognition of infants and young children (Malcolm Lewis in Harley and Woodward 1987).

Some reject all of those discussions out of hand and argue that to the degree that mapping is innate (that is, simply spatial cognition), it is shared by most of the animal kingdom. It is only humans that engage in mapmaking and only complex hierarchical cultures that store maps as part of an arsenal of coercive tools designed to reinforce particular aspects of the social structure. Thus, maps are sociocultural constructs

(Turnbull 1989; Rundstrom 1990). Denis Wood (1992) takes a relatively extreme position and argues that maps, as we know them, cannot be presumed to have existed in societies without the sort of complex hierarchical structure that characterizes our own. The argument is that, as a map is defined by the kinds of social relations to be found in our society, there is no relationship at all between a modern topographic map and the map-like cave drawings that date back to 40,000 years before the present. Were we to follow that line of argument, we would eliminate most branches of history that in any way trace today's ideas, technology, or science back beyond approximately the Renaissance. The argument that Wood makes is logical but anti-historical and, for most of us, absurdly counterintuitive. I would argue that maps (or, if you prefer, map-like depictions) are old, at the very least 40,000 years old; perhaps it is this age and the association of maps with our earliest cognitive development that give them their peculiar accessibility and fascination.

Maps have advantages over speech, gesture, mnemonic symbolism,[1] and dance. In spite of its extraordinary importance in human communication, speech, like writing, is essentially linear. One word follows another, and ideas about relationships that are not strictly linear are ill-served by the linearity of spoken and written language. Try describing the layout of a nearby city park and its use by different segments of the population: in no time you will find yourself gesticulating. The map must also have constituted a significant advance on gesture. If you wanted to save yourself the trouble of repeating an explanation, if you simply wanted to remember what you had said, or if your explanation of the use of the park became complex, you would soon find yourself sketching out a map with whatever tools were at hand. The map was also a remarkable advance on the mnemonic symbolism that characterizes much prehistoric rock art. Typical of such a mnemonic symbol is the spiral. Prehistorians believe that it was commonly used to commemorate an event. Because the spiral in no way sought to describe the event, movement from commemoration to description required the evolution of more complex symbolism. Folk tales and studies of nonliterate peoples have suggested that dance may have been an early way of expressing spatial relationships and of recounting events with a spatial dimension. The thought is delightful, but what an inconvenient and tiring form of expression for anything but ceremonial commemoration.

Maps, then, when they first began to be used must have brought a revolution in thought, perhaps associated with the very beginnings of descriptive science. They allowed for the expression of nonlinear spatial relationships and recorded these for reuse, for posterity, or for evidence. This in turn enabled increased clarity of spatial expression and simplified the conveyance of such spatial relationships. Some of this development still seems to be written, so to speak, on our modern maps in the symbolism we find there: mnemonic symbols that are arbitrary and designed to signify without describing; iconic symbols (reminiscent of early gestural communication and dance) such as a tree-like symbol to represent a tree or lined rectangles to represent cultivated fields; and conventional symbols, such as a dot to represent a household, which are themselves the product of a long history of map communication.

The Map and Geography

Historians of geography generally trace our discipline back to ancient Greece and sometimes back to ancient Egypt. In both cases, what is remembered as geographic in these societies is mapping: We look back through the works of Strabo and Herodotus to the maps of Anaximander, Aristagoras, the periplus (or navigational account) of the Carthaginian Hanno, and the calculations of Hipparchus and Eratosthenes. While the maps discussed by these ancient authors have long since disappeared, the maps in the Ptolemaic atlases (which are copies of copies of copies . . . vanishing back into the mists of time) reflect that ancient heritage. What is remembered in ancient Egypt is post-flood land surveying, maps of the regions of emerald mines, maps of the dead, and perhaps (depending upon whether we consider them to be maps or pictures) maps of the constellations. The cartography of ancient Egypt (prior to the Alexandrian conquest) is somehow of a very different world. What is remarkable about the cartography of ancient Greece, however, is how little the preoccupations, techniques, and even the awareness of the ideological potential of maps have changed since the days of Herodotus and Strabo (Jacob 1991). The contemporary preoccupation with knowing places also echoes ancient Greek geography; accounts of mytho-historic voyages, such as those of Homer and Agatharchide, were shaped by a cartographic vision of an imagined world.

Through much of the subsequent history of geography, until close to the end of the eighteenth century, geography and cartography were regarded as virtually synonymous. Indeed, there was no name to distinguish a cartographer from a geographer: the "cartographer," as distinct from the geographer, did not exist. Maps were simply one of the most effective means by which a geographer—for the most part confined to a drawing table and library in a European metropole—described the earth. In France those who became "géographes du roi" were invariably mapmakers, and it was the mapmaking and geodetic or astronomical geographers, too, who became members of the royal societies or who were considered scholars and scientists rather than merely popularizers and educators. Well into the late eighteenth century, geography was primarily concerned with the determination and graphic and verbal representation of location and position.

Sometime during the eighteenth century—in part as a result of advances in mathematics, astronomy, geodesy, and field mapping —the determination of location at small and medium scales or for geographic mapping purposes ceased to be an intellectual problem. That is, the problems that determining location had presented were understood to be technically solvable. What had once been a more or less unified field called geography, concerned with the description and representation of the earth, splintered into three new fields. What had already become known as *geodesy* retained the title of a science and the focus on the problem of determining location at increasingly large scales. What came to be called *cartography* was a technology that retained the determination of location as a central concern and a strong graphic element. *Geography* kept the name, but its realm became associated with written description and the analysis of the top layer of the earth and everything on and immediately above it. The field of geography, then, having essentially solved its central problem, began to assume a new shape and sought new preoccupations. Cartography, around which geography had been structured since ancient times, now seemed an outmoded and somewhat unacademic focus to the geographers of the early nineteenth century (Godlewska 1989).

What followed was a long period of quest for a new disciplinary heartland—a quest some would say has not yet been fulfilled. Some geographers sought to define geography as an introduction to all of the other sciences; some assigned geography an essentially literary function; still others located the field in the metaphor of exploration. There was work explicitly linking geography and education; geography and ethnography; geography and colonialism; geography and territorial administration; geography and botany; geography and earth science. . . . The field, as description, had been so all encompassing that potential directions seemed beyond number.

Much of the work carried out during and since the nineteenth century, however, has been shaped by the underlying metaphor of the topographical map and a search for mappable spatial relationships, in both the human and the physical realm. Only relatively recently have geographers truly broken the bounds of the map in searching, for example, for social relations, such as social relations of power, less susceptible to mapping but nevertheless important to explaining social phenomena. Indeed, arguably, in the years since World War II, human and physical geographers have moved significantly away from the map metaphor, from each other, and from a strong historical disciplinary identification, thereby increasingly identifying themselves first as scientists, social scientists, or humanities scholars and only secondarily as geographers. Although geographers have largely abandoned the map as a principal product of their labor and to some extent as a metaphor, the public and other disciplines have continued to associate geography and the map. For those looking at geography from beyond its borders, the determination of location and mappable spatial relationships define the field. Most geographers today, however, like most social scientists, claim the right to explore intellectual problems without being tied to a restrictive definition of their realm.

Major Conceptual Developments in Map History

Through this long historical and disciplinary evolution, the map itself has been far from static. It has evolved and changed in terms of the concepts shaping it, its form and medium of communication, and the nature of the data it encompasses and represents. Here I focus on the development of the map in the Western scientific tradition, largely because this is the tradition that has most influenced the form and nature of the map as we now know it. Remember, though, that Europeans were adept at adopting map-related ideas from peoples with whom they traded, such as the Chinese, who gave Europeans paper and the compass, and the Arabs, who gave us our numbering system and who developed and transmitted much else in mathematics, science, and navigation. Europeans also incorporated ideas from people that they conquered such as the Aztecs, who may well have provided Europeans not only with geographic information but with some cartographic symbols. Picking apart such influences is a worthwhile but mammoth task well beyond the scope and mission of this essay.

Conceptual breakthroughs have radically changed the look of maps from their earliest aspect as petroglyphs, or as the almost unrecognizable Peutinger map (which depicts the known world on a single stretch of parchment 6.75 meters long and only 34 centimeters wide), or as the highly pictorial and metaphorical T-in-O maps (Harley and Woodward 1987). One of the most important and earliest of these innovations was the concept of uniform scale. Yet the need for uniform scale and other innovations has varied over time; even today some maps, such as the famous map of the United States as perceived by a New Yorker, intentionally lack uniform scale. Appreciating the evolution of concepts underlying maps, therefore, requires keeping in mind the purpose for which any given map was produced, as this may have as much influence on its ultimate look as the state of the science behind it.

The concept of uniform scale probably developed early in the history of mapping, perhaps through the Babylonian, Egyptian, and Roman traditions of land surveying, cadastral mapping,[2] and the drafting of city and building plans. Still, even large-scale town plans like the Forma Urbis Romae, a plan of Rome dating from sometime between 31 B.C. and A.D. 192, did not have uniform scale. Comparing the surviving buildings with the map has shown that the map enlarged some buildings while shrinking others. Whether this was done accidentally or deliberately—as in the case of promotional nineteenth-century city views, county maps (Conzen 1984), and present-day tourist maps— has not yet been demonstrated. There is little question, however, that when the Romans centuriated for land holding and taxation purposes,[3] as in the cadaster of Orange (or Arausio) in the South of France, they sought to maintain a uniform scale within the tolerance of the instruments available to them.

The concept underlying this—that space can be treated homogeneously—is extremely important scientifically. The concept of uniform scale lies behind the idea of the uniform grid and reflects an objectification of space. Space is not necessarily how I alone perceive it or how the ruler of this realm conceives of it; it is an objective reality that everyone must negotiate with. Although some social scientists today emphasize the importance of relativity—in culture, economic value, or history —the concept of objectified and uniform space was fundamental to Euclidean, Cartesian, Newtonian, and Laplacian science, and it remains important to large-scale mapping and to much of our day-to-day living.

The understanding, development, and use of projections was an innovation almost as important as the development of uniform scale. It appears, however, to have been a much more recent innovation. Most of us accept that we live on a roughly spherical planet. Yet, when we make maps we draw them on a flat surface. We do this for a variety of reasons: to depict all of the earth on a globe at the scales we most often need would require a sphere of such proportions as to render it useless. A map, in contrast to a globe, gives a picture of the world at a glance. Nevertheless, the depiction of the spherical earth on flat paper clearly distorts the true shapes, areas, or distances that we are trying to depict. The smaller the scale of the map, the greater the visual distortion. Projection-related distortion, however, occurs even on large-scale maps. In general, projections are a means of representing the round surface of the earth on a flat medium by using systematic, and usually mathematically calculated, distortion. This distortion is mathematically calculated to define the relationship of the flat map to the round earth and allows us to take measurements from the map with minimal error.

In a sense the first step toward understanding and developing map projections was the realization that the earth was a sphere. We now think that the sphericity of the earth was first argued on a theoretical level as early as the sixth century B.C., that it became largely accepted among Greek scientists by the fifth century B.C., and that during the fourth century the concept was strengthened and elaborated with empirical proofs. The first projection was probably produced in the second century B.C. by Hipparchus (ca. 190–126 B.C.) when he tried to depict the sky on a flat map. Unfortunately, we know relatively little about that projection, except that it was probably stereographic. More than a hundred years later, Marinus of Tyre (fl. A.D. 100) and Ptolemy (ca. A.D. 90–168) produced what we have come to believe were the first terrestrial projections. Marinus created a rectangular projection, and Ptolemy corrected Marinus' projection and created two conic projections.

Ptolemy has a special place in the history of map projections and indeed in mapping altogether because he provided detailed instructions on the construction of his projections. His method of construction was graphic—as in, draw a line from point *a* to point *b*—rather than mathematical, thereby allowing cartographers to reproduce his projections for centuries without necessarily understanding the mathematics behind them. True mathematical projections had to wait until Edward Wright's explanation of the geometrically accurate and conformal mariner's projection, or what came to be called the Mercator projection, in 1599.

Today, the number and variety of map projections that can be generated mathematically and on computer with either subtle or major differences in distortion is virtually limitless (Snyder and Voxland 1989). The earth can be projected on a facial profile, on the shape of a hand, or on any shape at all. The projection that is chosen, however, for the depiction of a map, such as one of the world, is of considerable importance as the projection gives the overall sense of the world in which we live and the very particular sense of our place within it. The Mercator map, probably still the most commonly known and used projection, distorts the higher latitudes and makes the most populated and non-European parts of the world seem insignificant in the grand scheme of things (Figure 1.2). The Robinson projection diminishes the exagger-

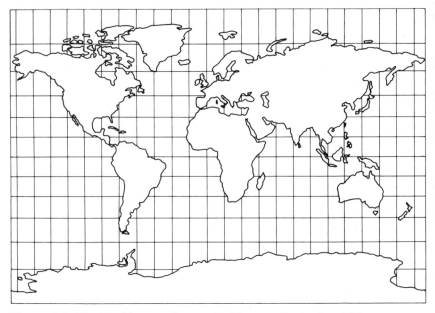

Figure 1.2. The World, according to the Mercator Projection. This projection, originally developed in the sixteenth century to facilitate transAtlantic oceanic transportation, became the chosen projection of nineteenth-century school atlases. It exaggerates the Northern Hemisphere and shrinks Africa. Drawn by Ross Hough of the Queen's University Cartographic Laboratory.

ated size of Greenland to be found on the Mercator map but further shrinks an already small Africa (Figure 1.3). The Peter's projection, not so commonly seen in Western world maps, is an equal-area projection that does not exaggerate the size of any part of the earth (Figure 1.4). It has, consequently, come to be seen in many circles as a nonimperialist, nonracist projection and an appropriate cartographic structure for a world to be shared equally by all of its citizens (Wood 1992). Projection, then, is a vitally important concept that has had a considerable impact on the sciences (particularly on geography), on the arts (including the stimulation it may have provided to perspective theory), and on the social construction of weighted space or zones of importance —the very obverse, if you like, of the concept of objectified and uniform space.

One of the most conceptually significant and yet least written about aspects of the map is symbology. Cartographic symbology can be analyzed on at least three levels: the cognitive and conceptual significance of symbology, the complex layers of meaning expressed by

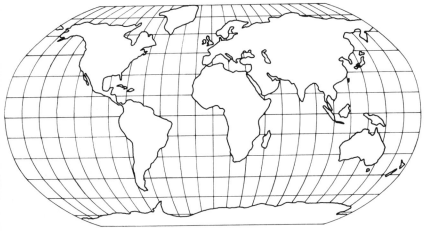

Figure 1.3. The World, according to the Robinson Projection (an elegant projection developed by Arthur Robinson). Drawn by Ross Hough of the Queen's University Cartographic Laboratory.

individual symbols and symbol types, and the evolution over time of symbols. In effect, everything on a map is a symbol including the title of the map, the neat line (the box around the map), the lettering, the river symbols—all are symbolic representations of a perceived reality, whether physical, cultural, imagined, or hallucinated. Nothing on a map is reality; everything is representation and, thus, open to the sort of exegetic and interpretive analysis that characterizes the humanities (Harley and Blakemore 1980).

The very act of symbolizing suggests a great deal about humans and their basic intellectual needs. Symbolic representation is analogous to the metaphor and to analogy in verbal and written communication with all their subtlety and power to evoke. Although we persist in seeing ourselves as rational computing creatures, balancing off costs and benefits with a full array of facts and evidence, every thinking person knows that the human mind functions by flitting about in search of meaningful patterns without a great deal of concern for data and evidence. In fact, we live in a hopelessly complex world where the data, if we truly sought all of it, would overwhelm us instantly and repeatedly.

The complexity of the world has always awed us, as the history of religion attests. Symbols allow us to reconstruct the complex world in simplified form, imposing an organizational schema on it and creating a whole series of visual and intellectual hierarchies. We are so needful of such intellectual constructions that we are unable to imagine a world without them, that is, without an imposed order and hierarchy

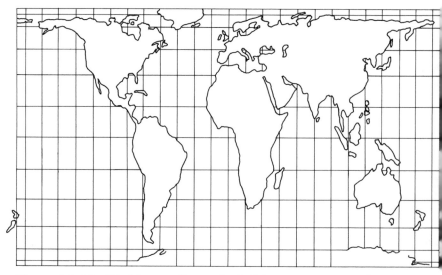

Figure 1.4. The World, according to the Gall Projection (an equal-area projection that is far earlier and less politically charged, but no less effective, than the Peters Projection). Drawn by Ross Hough of the Queen's University Cartographic Laboratory.

of some sort. Indeed, we even have difficulty communicating between different imposed orders and hierarchies—such as between Muslims and Christians, Laplacians and Cartesians, and round and flat earthers. The use and recording of symbols, however, also permit the rapid transmission of information and, as Christian Jacob (1992) points out, "the confrontation of experience." Symbols have the potential to help us confront and ultimately reconcile conflicting models of the world. A map and its system of symbols is a simplification of reality that reflects a particular understanding of the world and that can contribute to subsequent or competing models of reality.

The symbols used by cartographers have many layers of meaning (Bertin 1983; Harley 1982; Tufte 1983). The neat line, for example, serves to focus the attention of the reader, but, depending on its form, it can also separate cartographic from pictorial space (as when the neat line is ringed by pictorial images) or it can imply the relatedness of all representations— whether cartographic, pictorial, or textual— within its borders. It can suggest the scientific reliability of the map and mark out the limits of the known world. It may convey a sense of the distance between the map reader and the area being depicted (as when the border simulates a picture frame or a doorway). It can subtly reinforce, by the inclusion or exclusion of key territory, a particular territorial sovereignty. Or by the exclusion of a territory, it may imply the very biased nature of a given depiction. A neat line can carry a wealth of meaning often absorbed only unconsciously by most map readers.

Similarly, a map with a variety of town symbols composed of castle walls, churches, and a jumble of houses can give the often faulty impression that the cartographer has personally visited each town. On the other hand, a map that depicts all towns with dots of varying size creates the equally faulty impression that all towns of a given size are fundamentally the same. Another extremely important symbol, the map title, can be taken at face value as simply reflecting what is being depicted. The words "nation," "realm," "country," "territory," however, all carry different connotations and, in a sense, pre-shape the attitude of the reader to the subject. Similarly, the use of a long-forgotten country name in the title of a map may not just be odd or quaint, but speak of a conquering power's greater identification with the region's history than with its present inhabitants. Even the depiction of physical features can reflect values that have little to do with the existence of the features in nature. Thus, because the scientific imagination of the time sought regularity, repetition, and predictability, many late-eighteenth- and early-nineteenth-century maps depicted a regular and almost geometrical pattern of mountains and river systems. Every single feature on a map will reveal multiple levels of meaning and surprising preconceptions, right down to the coastal outline, which seems to claim that there is indeed a place where water and land are entirely separate and distinct.

Cartographic symbolism has not been static through the ages. Mountain symbols, for example, have transmogrified from sugar loafs to curved lines depicting humps and collections of humps, to a system of hachures illuminated from the northwest,[4] to shadow effects without hachures, to contours that look like waves resulting from a pebble dropped into a pool, and to all sorts of combinations of these (Robinson 1982). The history of their formation and the other symbols to be found on a map is the history of the development of a cartographic language or several languages or a series of languages with different dialects. Few attempts have been made to document the development of these languages or even of parts of them. The work of Francois de Dainville (1964), who traced the definitions of geographic terms and map symbols as far back as the tenth century for some symbols, is truly exceptional. Through such reconstructions we can begin to answer larger questions such as to what degree changes in cartographic

language reflect changes in understanding of the phenomena themselves or changes in the technology of data collection, depiction, or
reproduction. Or, what did the decreed uniformity of cartographic
symbolism in France in the early nineteenth century mean and what
were its consequences for cartography, for graphic communication,
and for geo-knowledge?

Major Technical Developments in Map History

Conceptual developments have not been the only source of innovation
in map history. Revolutions in the technology of reproduction have
also changed the shape and impact of the map as have changes in data-
collection technology. As I noted earlier, an extraordinary variety of
materials can and have been used as a cartographic base. The development of rag paper—an innovation we think emanated from China
sometime around the eleventh century A.D.—revolutionized written
and graphic expression. Paper's advantages were numerous: it was
cheaper and easier to make than parchment, more portable than stone
and metal alloys, more available than papyrus, and if preserved from
moisture and sunlight, reasonably long-lasting. Certainly paper, and
especially the innovation of pulp paper in the nineteenth century,
made documents, drawings, and books affordable to a significantly
larger population.

Whereas the innovation of paper prepared the way for widespread
diffusion of written or graphic material, it was the printing press and
printing technology in general that over a period of five centuries (beginning in the fifteenth century) made maps, etchings, engravings, and
texts available to virtually everyone (Woodward 1975). The availability
of particular maps or particular types of spatial information to the
general public (such as, for example, the location of nuclear dump
sites, the position of strategic defensive or offensive weapons, and any
other information that the powerful deem legitimately or illegitimately
to be "sensitive") will, however, likely always be restricted. The popu-
larization of written and graphic media is an important chapter in human history.

In combination, paper and printing also made possible the building
of shared cultural traditions covering increasingly larger territories together with the possibility of prolonged intellectual exchange. In
cartography, these innovations allowed the accumulation and constant correction and updating of geographic knowledge—a geographic
knowledge that some believe contributed to the European domination
first of its own territory and then of much of the rest of the world.

Figure 1.5. This Turkish Airlines advertisement uses three languages—the map, the photograph, and the text—to express the airline's technological embodiment of the historical, cultural, and geographical centrality of Istanbul.

Indeed, cartographic culture today is no longer merely regional, national, or even continental, but global. This scope is clear in the embodiment of the globe and the map as symbols of international economic power: Witness the frequent cartographic motifs to be found in the advertisements, articles, and front covers of the magazine specializing in international economic affairs, the *Economist* (Figure 1.5). This cartographic globalization is, however, only part of the picture as global, national, and even regional control has long been and still is contested, perhaps most effectively when the same cartographic tools are used in the fight for local autonomy.

Data collection techniques have acquired a similar sophistication, comprehensiveness, and scope. Early in European history there was a limited ability to collect reasonably accurate latitudinal locational information and locational information over small areas. From the eighteenth century, data collection techniques aided by the Harrison chronometer and instruments such as theodolites, plane tables, telescopes, and alloy chains made it possible to collect longitudinal positions and information on more extensive areas with some accuracy. Since then, the technology and mathematics in this field, known as geodesy, has evolved to the point where it may be possible to program a missile to travel a few feet above the surface of the earth from a specific address on one side of the globe to a specific address on the opposite side of the globe. Indeed, the accuracy and detail of cartographic information now available through, for example, satellites and laser-ranging technology, in combination with the storage and retrieval capacity of computers, have exploded the borders of the paper map, reduced the cost and time required to produce it, and increased the distribution efficiency. These developments have moved a significant sector of the mapping business from compilation and graphic reproduction to the management of bits of information. An increasing demand for positional accuracy has generated only part of this explosion of cartographic information. Equally important has been the recognition of the value of a variety of types of spatial information, particularly in planning for the provision of social services such as medical facilities services, sewerage, electrical systems, and transportation—again on a local, regional, national, and even global scale.

The Power of Maps

Throughout their history, whether as rock carvings, paper documents, or computer-generated images, maps have had remarkable power.

They are documents that can draw a crowd, be it composed of art specialists or military strategists. What is the source of this power?

The manner in which the map conveys information is one of the most important sources of the map's power to influence. As already noted, in contrast to texts, language, gesture, and even to algebra, maps are read situationally rather than linearly; that is, they are read as we tend to perceive the world around us. To some extent, you will look at a map in much the same way that you look at the countryside from atop a tall building. The landscape and map reader will immediately seek orientation, scale, familiar symbols or landmarks, recognition of location, and any connection with the experience or the knowledge of the reader. Even the pattern of eye movement will be similar, as the reader's eyes dance about the map or the landscape. That is undoubtedly a good part of what makes maps so compelling to the average person. Texts must be opened and read according to forms and procedures taught over long years. A map speaks more directly and simply. Good map reading requires knowledge and skill, but maps will speak to many for whom texts remain mute.

Perhaps in part as a result of this, maps are more universally accessible across languages and cultures. It is not difficult to pick out the basic elements of most maps with or without knowledge of the relevant language—although complete ignorance of the culture that produced the map and its purpose may mislead. In addition, as pictorial images, maps can tell a story that leaves much to the imagination and to interpretation; when the map reader and the mapmaker come from different cultures, maps can lead to miscommunication as the reader attempts to bridge the gap and to smooth over what is not shared between the cultures.

The effectiveness with which maps simplify predigested information and give it hierarchical form gives maps much of their strategic utility. The culmination of detailed research into the spatial characteristics of phenomena, maps display their information with remarkable simplicity and clarity. Political and military strategists may frequently think in the abstract, but they usually are obliged to plan and act in concrete and Euclidean space, or risk seeing their plans go unrealized. As military strategists through the ages have appreciated, hardly anything is more useful than the appropriate map at the appropriate scale at the appropriate time. Further, it is, as any corporate executive will confirm, predigested information that allows speedy and efficient decision making. To be powerful, the picture must be simple. Visual simplicity does not mean, however, that the information within the map need lack complexity and depth. The map is an unusual document, bearing a great deal of difficult-to-gather information, which nevertheless presents that information with a combined pictorial simplicity and complex symbolic resonance. This resonance is significant enough that, were we methodically and critically to disassemble the map, many much more complex levels of meaning would be revealed—levels the strategist might have been unaware of. Yet who is to say how the map in toto influenced action?

The precision of the lines on the map, the consistency with which symbols are used, the grid and/or projection system, the apparent certainty with which place names are written and placed, and the legend and scale information all give the map an aura of scientific accuracy and objectivity. Although subjective interpretation goes into the construction of these cartographic elements, the finished map appears to express an authoritative truth about the world, separate from any interests and influences. The very trust that this apparent objectivity inspires is what makes maps such potent carriers of ideology (Harley 1989). However unobtrusively, maps do indeed reflect the world views of either their makers or, more probably, the patrons of their makers, in addition to the political and social conditions under which they were made. Some of the simple ideological messages that maps can convey include: This territory is and has long been ours; here is the center of the universe; what counts is not the people but the state; territorial conquest is a glorious and righteous mission; if we do not claim this land, the enemies you most fear will.

A portion of the power of maps comes from their long association with power. Maps have generally expressed knowledge about places from the vantage point of an elite. The production of maps—certainly the useful ones—demands, and seems always to have demanded, considerable outlays of capital in the form of training, time, field and archival research, materials, and even health and lives. Few indeed have been the merchants and leaders either able to afford such an investment or possessed of the imagination to understand its value. The information to be found on the most useful maps produced through history was, consequently, privileged, restricted, and thus imbued with social and political power.

Conclusion

Maps have been, and continue to be, among the most powerful of geographic ideas. Often truly multimedia and almost as diverse in structure as space itself, the map has been used for an extraordinary variety of purposes, from the strategy of warfare and business, to the exploration of human physiology, to deceptive persuasion. Shrouded in mystery, the origins of the map suggest something of the early history of human cognition, language, and religion. Although the map predates geography, geography from its earliest history so effectively structured itself around the map and the solution of spatial problems that geography and the map became inseparable in the popular imagination. Over the course of its history, the map itself has undergone considerable change from a document more like art work in its depiction of highly individuated subjective space to a representation of uniform space with an explicit recognition of the distortion to be found in flat maps and a constantly changing symbolic system of expression. Its importance over time has also vastly increased as the materials and technology of its making have rendered it virtually omnipresent. Indeed the map has become so influential that it has at once become a taken-for-granted daily companion and a subtle yet powerful shaper of our understanding of the world, our territory, each other, and our past.

Notes

1. Mnemonic symbols, often very simple, are designed to aid in the memory of (or to commemorate) an event.
2. Cadastral mapping is mapping for land registration and taxation.
3. Centuriation is a surveying process that divided specified lands into 100 plots to match the military units of one hundred men to be settled there.
4. Hachuring is a technique of relief depiction composed of a series of short thin parallel lines, generally running in the direction of rainfall, and most dense at points of greatest slope.

References

Bertin, Jacques. 1983. *Semiology of Graphics: Diagrams, Networks and Maps.* Madison: University of Wisconsin Press.

Castner, Henry W. 1990. *Seeking New Horizons: a Perceptual Approach to Geographic Education.* Montreal, Kingston: McGill University Press.

Conzen, Michael P. 1984. The County Landownership Map in America: its Commercial Development and Social Transformation, 1814–1939. *Imago Mundi* 36:9–31.

Dainville, François de. 1964. *Le langage des géographes.* Paris: Editions A. et J. Picard et Cie.

Godlewska, Anne. 1989. Traditions, Crisis, and New Paradigms in the Rise of the Modern French Discipline of Geography 1760–1850. *Annals of the Association of American Geographers* 79(2):192–213.

———. 1994. Napoleon's Geographers: Imperialists and Soldiers of Modernity. In *Geography and Empire*, ed. A. Godlewska and N. Smith, 31–53. Oxford: Blackwell.

———. 1995. Map, Text and Image. The Mentality of Enlightened Conquerors: A New Look at the *Description de l'Egypte*. *Transactions of the Institute of British Geographers* 20(ns):5–28.

Gould, Peter. 1974. *Mental Maps*. Harmondsworth: Penguin.

Hall, Stephen S. 1992. *Mapping in the Next Millennium: the Discovery of New Geographies*. New York: Random House.

Harley, J. Brian. 1982. Meaning and Ambiguity in Tudor Cartography. In *English Mapmaking 1500–1650: Historical Essays*, ed. Sarah Tyacke, 22–45. London: British Library Reference Division Publications.

———. 1989. Deconstructing the Map. *Cartographica* 26(2):1–20.

Harley, J. Brian, and Michael J. Blakemore. 1980. Concepts in the History of Cartography: A Review and Perspective. *Cartographica Monograph* 26.

Harley, J. Brian, and David Woodward. eds. 1987. *The History of Cartography*, vol. 1, *Cartography in Prehistoric, Ancient and Medieval Europe and the Mediterranean*. Chicago: University of Chicago Press.

Hewes, Gordon, W. 1977. A Model for Language Evolution. *Sign Language Studies* 15:97–168.

Jacob, Christian. 1991. *Géographie et ethnographie en Grèce ancienne*. Paris: Armand Colin.

———. 1992. *L'Empire des cartes. Approche théorique de la cartographie à travers l'histoire*. Paris: Albin Michel.

Kendon, Adam. 1975. Gesticulation, Speech and the Gesture Theory of Language Origins. *Sign Language Studies* 9:349–373.

———. 1981. Geography of Gesture. *Semiotica* 37(1–2):129–163.

Lynch, Kevin. 1960. *The Image of the City*. Cambridge: MIT Press.

McGuinness, Carol. 1992. Spatial Models in the Mind. Special Issue: Perceptual Constancies. *Irish Journal of Psychology* 13(4): 524–535.

McNeill, David. 1992. *Hand and Mind: What Gestures Reveal about Thought*. Chicago: University of Chicago Press.

Meggitt, M. 1954. Sign Language among the Walbiri of Central Australia. *Oceania* 25:2–16.

Munn, Nancy. 1966. Visual Categories: An Approach to the Study of Representational Symbols. *American Anthropologist* 68(4): 936–951.

Robinson, Arthur H. 1982. *Early Thematic Mapping in the History of Cartography*. Chicago: University of Chicago Press.

Rundstrom, Robert. 1990. A Cultural Interpretation of the Inuit Map of Accuracy. *Geographical Review* 80:155–168.

Snyder, John P., and Philip M. Voxland. 1989. An Album of Map Projections. *USGS Professional Paper* 1453. Washington, D.C.: U.S. Government Printing Office.

Tufte, Edward R. 1983. *The Visual Display of Quantitative Information*. Cheshire, Conn.: Graphics Press.

Turnbull, David. 1989. *Maps Are Territories, Science Is an Atlas: Portfolio of Exhibits*. Geelong, Australia: Deakin University.

Von Raffler-Engel, Walburga, Jan Wind, and Abraham Jonker, eds. 1991. *Studies in Language Origins*, vol 2. Amsterdam and Philadelphia: Benjamins.

Wind, Jan, Edwin G. Pulleyblank, Eric de Grolier, and Bernard H. Bichakjian, eds. 1989. *Studies in Language Origins*, vol 1. Philadelphia: Benjamins.

Wood, Denis, with John Fels. 1992. *The Power of Maps*. New York: Guilford Press.

Woodward, David, ed. 1975. *Five Centuries of Map Printing*. Chicago: University of Chicago Press.

———. 1987. *Art and Cartography: Six Historical Essays*. Chicago: University of Chicago Press.

2

The Weather Map: Exploiting Electronic Telecommunications to Forecast the Geography of the Atmosphere

Mark Monmonier

Weather is the soap opera we all watch. Each day's weather leaves us in suspense about the next, and every day adds its twist to the season's plot. An ongoing drama that affects our lives, the weather story is common ground for casual conversations with intimate friends and new acquaintances. What we rarely talk about, though, is the way the story reaches us. Instead of going outside to sniff the breeze, we merely turn on the television or check the daily paper for a wealth of information about today's weather, tomorrow's weather, and weather across the country. Weather maps have well-established slots on morning and evening TV newscasts and are the starring attraction on their own cable channel, available 24 hours a day. Early risers who catch the aviation weather briefing on public broadcasting receive an in-depth assessment of how today's and tomorrow's weather will affect airports and airlines, and on evening newscasts at six and eleven, the local meteorologist who recaps national and local weather uses a regional radar map to point out intensely falling rain or snow. In addition, the "satellite loop" that compresses space and time describes the jet stream's mood swings and the birth and death of storms. Snapshot views of the state of weather are so readily available, we easily forget they are one of the great inventions of modern geography.

Imagine what life was like a century ago, when weather maps were bland and comparatively scarce, and two centuries ago, when they didn't exist at all. In the late 1790s, farmers, mariners, and generals scanned the sky for clues that fit weather proverbs such as "red sky at morning, sailor take warning!" Although weather lore was useful in forecasting rain, these vague aphorisms were less accurate than modern forecasts and revealed little about amount and duration of precipitation. Scholars had yet to discover reliable forecasts, which required synchronized measurements of temperature, pressure, wind, and sky conditions taken almost simultaneously at dispersed locations linked by a well-coordinated network for rapid data collection and timely mapping and interpretation. Discovery of the value of weather maps in the 1820s and 1830s, together with the development of electrical telegraphy in the 1840s and 1850s, spurred investment in the costly infrastructure needed to collect synchronous weather data.

This chapter examines the evolution of the weather map and its role in predicting weather and understanding atmospheric processes. As an adaptation of an earlier geographic idea—the idea of the map, described in Chapter 1—the weather map organized atmospheric data and promoted the empirical study of weather phenomena and a theoretical understanding of storms and atmospheric circulation. Mapping weather was a natural consequence of eighteenth-century refinements of the thermometer and the barometer. In the early nineteenth century, experimental charts based on weather data several decades old revealed the potential of the map as an effective forecasting tool, but only after 1870 did the idea of mapping the weather overcome the technological and political obstacles to the formation of a national weather service.

More an institutional breakthrough than a cartographic innovation, the nineteenth-century weather map raised issues that are still the focus of debate, including the roles of private enterprise, the use of maps for propaganda, and the effect of budgeted appropriations on data quality. Yet the form and function of weather maps were not simply dictated by government policy. Present-day influences on their format and use include satellite remote sensing, lively weather graphics in the mass media, computer forecasting models, automated sensors for monitoring sky conditions at airports, and scientific visualization. Equally important, the modern weather map encourages an increased appreciation of our connections with distant places, including links between their weather and our weather, as well as a recognition of our ability to prepare for weather, even though we can't control it.

A Cartographic Snapshot of the Atmosphere

The weather map is a cartographic snapshot of the atmosphere. A single display integrates information about several important weather elements for a stated instant of time, such as 8 A.M. eastern standard time. Some weather elements, such as temperature and barometric pressure, must be measured with carefully calibrated instruments, whereas others, such as the presence and type of precipitation, are merely observed. Still others, such as the degree of cloudiness (clear, partly sunny, partly cloudy, or cloudy), require standard criteria so that observers in different locations can report sky conditions reliably and consistently.

The most remarkable weather map is the forecast map, which describes the likely state of the atmosphere at a specified time in the near future. The forecast map helps warn state and local authorities, farmers, shippers, and pilots about hurricanes, tornadoes, blizzards, cold waves, and other severe weather conditions. Storm warnings are important because timely evacuations and preparations can save lives and protect property. On television and in daily newspapers, the forecast weather map helps to explain local forecasts, which citizens use to plan picnics, yard work, and house painting.

The weather map has made forecasting possible in two ways. In a direct sense, the weather snapshot is a starting point for estimating the state of the atmosphere a half day, a full day, or several days hence. In an indirect sense, maps of current weather have helped scientists discover atmospheric processes and acquire the knowledge that makes weather forecasting more than a guessing game or a crapshoot.

One particularly useful cartographic revelation is the relationship between pressure and wind. Figure 2.1 is a schematic weather map showing the circulation of air inward toward the center of a low-pressure cell labeled L. A concentric pattern of *isobars*, or lines representing equal barometric pressure, describes a "depression" in the atmospheric pressure surface. Surface winds develop as air flows from areas of higher pressure on the periphery of the storm to areas of lower pressure closer to the storm's center. Deflected to the right by the coriolis effect,[1] the winds move inward in a counterclockwise, "cyclonic" pattern, described by the arrows. Relatively close isobars, as in the area labeled A, indicate zones with comparatively large differences in pressure over a short distance—these steep "pressure gradients" can cause particularly severe, destructive winds. Deep depressions reflecting extreme contrasts in pressure between center and edge generally produce the most severe storms.

Knowledge of storms and a consecutive series of current weather maps help the meteorologist predict where a storm will go and what effects it will have. In North America and Europe, for instance, low-pressure cells and other weather systems generally move from west to

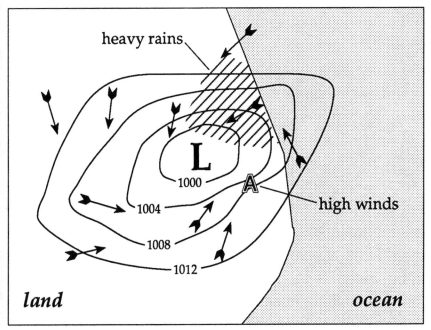

Figure 2.1. Hypothetical weather map illustrating the counterclockwise winds around a low-pressure cell and a belt of heavy precipitation (shaded area) between the storm center and a supply of moisture.

east.[2] A forecaster tracking a pressure cell cartographically can establish the storm's path, make an educated guess about its future position, and issue appropriate warnings. Because weather maps describe not only the size and intensity of the storm but also the supply of moist air, the forecaster can also estimate the amount and duration of precipitation. In Figure 2.1 the shaded gray area to the northeast of the storm center is a belt of intense precipitation lying between the center of the low-pressure cell, toward which the air is moving, and an ocean supplying abundant moisture through evaporation. A coastal storm moving from southwest to northeast typically has two principal phases. As the storm enters the area, pressure drops and heavy rain accompanies northeast winds. Later, as the storm moves on, pressure rises and clear skies accompany southwest winds. Although weather proverbs might describe this sequence, they are no substitute for timely warnings based on cartographically organized, systematically collected weather data.

Weather Mapping Prior to 1870

Forecasts based on weather maps faced three obstacles: the absence of instruments to measure the atmosphere, insufficient data to make the maps informative, and the lack of a network for collecting timely data. Although scientists and natural philosophers could measure temperature and pressure by 1800, the value of mapping the instantaneous state of the atmosphere was not immediately obvious. To demonstrate the efficacy of the weather map as a forecasting tool, they first had to overcome distance. Although a slow, informal network would suffice to prove weather maps were scientifically revealing, much faster communication was needed to show that forecasting could be publicly beneficial.

Like most early scientific instruments, the thermometer and barometer passed through several stages of refinement. Although Galileo had demonstrated a crude "thermoscope" as early as 1593, temperature measurement was inaccurate until around 1650, when Leopoldo, Cardinal dei Medici, devised a crudely calibrated, closed-glass thermometer by attaching a linear temperature scale to a narrow glass tube closed at the top and with a chamber of wine at the bottom. In 1714, Gabriel Daniel Fahrenheit developed a simple mercury thermometer calibrated to 32° (the freezing point of water) and 96° (the normal temperature of the human body), and in 1742, Anders Celsius developed the centigrade (0°–100°) scale anchored to the freezing and boiling of water at sea level. The first barometers exploited similar materials: In 1643, Evangelista Torricelli demonstrated the principle of atmospheric pressure by crafting a glass tube slightly more than 30 inches long and closed on one end, filling the tube with mercury, anchoring it to a vertical ruler, and inserting the open end in a cup of mercury. In 1843, in France, Lucien Vidie invented the less cumbersome, less expensive aneroid barometer, which provided direct readings with a pointer attached to a thin-walled, pressure-sensitive chamber.

In the late eighteenth century a number of European scientific observers began to collect and share weather information. The Societas Meteorological Palatina, founded in 1780, encouraged its members to record systematic observations of temperature, pressure, wind direction, precipitation, and sky conditions. Around 1820, H. W. Brandes, professor of mathematics at Breslau, used these data to compile and publish a series of synchronous daily weather maps for 1783 (Robinson 1982:73–74). Brandes's map for March 6, 1783, believed to be the first weather map, demonstrated the effect of atmospheric pressure on wind direction and speed during a severe storm that caused considerable destruction in northern Europe. As in Figure 2.1, isobars outlined the storm's low-pressure cell and revealed a marked correlation between pressure gradient and wind.

During the next three decades, weather mapping in Europe and North America led to widespread recognition of the cyclonic nature of storms and the feasibility of meteorological forecasting based on timely weather data. Scientists recognized the importance of observations recorded within an hour or less of each other and the value of organizing these data on a composite plot showing temperature, pressure, wind, and areas of precipitation. In the 1830s, controversy about the behavior of storms led to personal networks of observers, sometimes with institutional support (Fleming 1990:23–25, 40, 63). William Redfield, a leader of the "Boston circle" of weather observers, advocated a kinematic "whirlwind" theory based on gravity and the rotation of the earth. He corresponded with a loose network of scientists and weather observers in America, Europe, and the Caribbean. James Espy, a leader of the "Philadelphia circle," advocated a theory emphasizing heat and convective flow. Recognizing the limitations of his personal network, Espy developed various institutional networks during the 1840s and 1850s with the support of the U.S. Navy, the Army Medical Department, and the Smithsonian Institution. By refining and publicizing meteorological concepts, these theoretical debates based on informal networks nurtured the development of a forecasting network.

Sparse data-collection networks focused attention on major storms of sufficient intensity, breadth, and duration to reveal the strongest, most apparent geographic patterns. In 1843, Elias Loomis (1846), a professor of mathematics and natural philosophy at Western Reserve College, in Cleveland, Ohio, addressed the American Philosophical Society with a seminal paper titled "On Two Storms which Were Experienced Throughout the United States, in the Month of February, 1842." Loomis mapped these storms using barometric and other weather data from 68 registers, some provided by Espy, and additional information on wind, temperature, and precipitation from 41 military posts and 22 other observers. Figure 2.2 is the third of five charts Loomis used to describe the development and movement of a severe winter storm. The dashed lines are deviation isobars, with the line labeled 0 representing mean barometric pressure and those labeled $+2$, -2, and -4 representing pressures two inches higher and two and four inches lower than normal, respectively. (Deviation isobars adjust for the effect of elevation on pressure. A similar scheme portrays temperature deviations with coarse, numbered dotted lines.) Within the -4 isobar a north-south dashed line labeled -6 marks the center of the low-pressure cell. Loomis observed that to the west of this line the

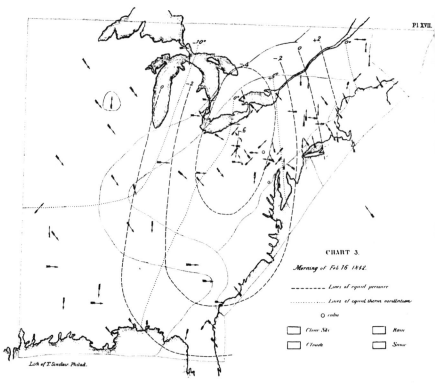

Figure 2.2. Loomis's third chart for the storm of February 15–17, 1842. The thin, unnumbered dotted lines are guidelines for adding hand-colored area symbols for the four sky-condition and precipitation categories in the map key. Source: Loomis (1846, plate xvii).

winds were blowing largely from the west or northwest whereas to the east of the line the wind tended to be from the south or southeast, suggesting a counterclockwise flow. Successive maps traced the movement of the storm center from the Ohio Valley on the morning of February 15th to a position over Atlantic Canada on the morning of February 17th. Loomis's paper was the first publication of a synoptic weather map using isobars and isotherms (Miller 1933:191).

At the conclusion of his paper, Loomis offered a highly optimistic assessment of the use of weather maps as forecasting tools:

> If we could be furnished with two meteorological charts of the United States, daily, for one year, charts showing the state of the barometer, thermometer, winds, sky, &c., for every part of the country, it would settle for ever [sic] the laws of storms. No false theory could stand against such an

array of testimony. Such a set of maps would be worth more than all which has been hitherto done in meteorology. Moreover, the subject would be well nigh exhausted. But one year's observation would be needed; the storms of one year are probably but a repetition of those of the preceding. (Loomis 1846:183)

Despite his naiveté about the periodicity of weather, Loomis was well aware of the need for a vastly improved sampling network:

> We need observers spread over the entire country at distances from each other not less than fifty miles. This would require five or six hundred observers for the United States. About half this number of registers are now kept, in one shape or another, and the number, by suitable efforts, might probably be doubled. Supervision is needed to introduce uniformity throughout, and to render some of the registers more complete. Is not such an enterprise worthy of the American Philosophical Society? (Loomis 1846:184)

The inability to collect data rapidly and efficiently was clearly the major impediment to forecasting weather. As early as 1793, the French chemist Lavoisier had backed Claude Chappe's scheme for a visual aerial telegraph, which would transmit messages in graphic code across a network of semaphore towers. But Lavoisier was a casualty of the French Revolution, and Chappe's labor-intensive visual telegraph, although tested, was never extensively developed. In 1842, five years after the invention of the Morse telegraph and coincident with the successful demonstration of a prototype system linking Baltimore and Washington, Prague scientist Carl Kriel proposed using the electromagnetic telegraph to compile timely weather maps. In 1848, John Bell presented a similar suggestion to the British Association for the Advancement of Science. In 1858, the French astronomer Urbain Le Verrier began using telegraphed data to make synoptic weather maps. He won the support of Napoleon III in expanding his network of observers throughout much of Europe by convincing the emperor that timely weather information during the Crimean War might have avoided the destruction of an allied British-French supply fleet by a severe storm (Miller 1933). Although the Paris Observatory's *Bulletin Internationale* had carried weather data collected by telegraph since 1858, the *Bulletin* did not include timely weather maps until September 16, 1863. Nonetheless, Harrington (1894:330) credits Le Verrier with being the first to publish a current weather map based on data collected the same day.

The American telegraph network developed so rapidly that by 1846 Redfield, of the Boston circle, pointed out the practicability of transmitting timely warnings of the approach of West Indian hurricanes.

Joseph Henry, who became the first director of the Smithsonian Institution in 1846, joined with Espy in recruiting a network that numbered 150 observers in 1849 and eventually grew to 500 observers. The Smithsonian network began to collect telegraphic observations in 1857, but in 1861 the Civil War severed Washington from many places to the west and south, placed a heavy load on remaining links, and prematurely ended Henry's innovative experiment. Although he had demonstrated the practicability of weather telegraphy early in the war, Henry was unable to convince either Congress or the Smithsonian's board of regents to back his proposal for a weather service similar to the network Le Verrier had extended across much of Europe (Miller 1933:192). As the end of the war approached, Henry renewed his call for a national weather service, but in January 1865 a fire destroyed much of the Smithsonian building, disrupting the meteorological program and diverting Henry's attention for the next several years. Although a self-published history of the Smithsonian Institution makes a plausible case that Henry's network of telegraphic observers was the "beginnings of the Weather Bureau" (True 1929), in the post-war years the Smithsonian's meteorological project was too poorly funded and too disorganized to be an effective catalyst (Fleming 1990:148–150).

In Britain, a more conveniently compact country, the press played an important role in early efforts to collect timely weather information. In 1848, London's *Daily News*, for instance, published the first telegraphic weather report. In 1849, the *Daily News* began publishing a table of meteorological observations, mostly carried from outlying points by train. For two months during 1851, at the Great Exhibition in Hyde Park, entrepreneur James Glaisher printed a daily weather map based on information collected the same day. Around 1861, the Daily Weather Map Company, a London firm, compiled and sold a daily weather map (Marriott 1901, 1903). As Figure 2.3 illustrates, the map was not a multi-theme synoptic map with isobars and isotherms, but a geographic array of pictograms describing weather conditions at selected cities in Britain and Ireland. Government became more fully involved when Admiral FitzRoy organized weather telegraphy around 1860, and the newly established British Meteorological Office began making daily weather charts in 1867. The press was pleased to obtain its weather data from the government, and government was pleased to have the press disseminate its forecasts and warnings. On April 1, 1875, the *London Times* became the first newspaper to carry a daily weather map, a generalized, column-width version of the national weather map prepared by the Meteorological Office but paid for by the newspaper (Scott 1875).

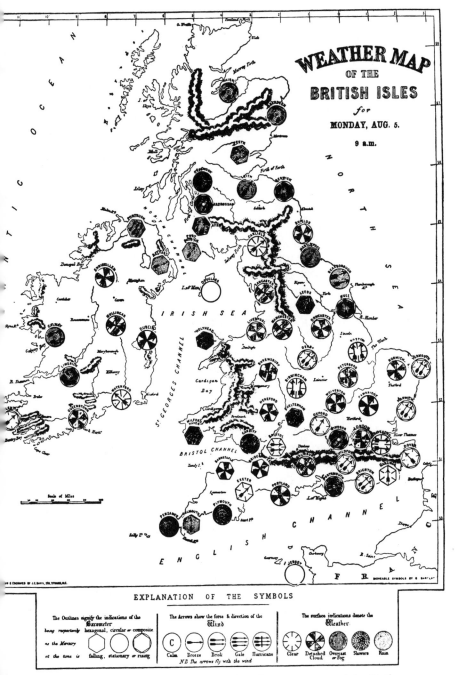

Figure 2.3. Sample weather map accompanying the prospectus of the Daily Weather Map Company, around 1861. Source: Marriott (1901) and Shaw (1926:309).

Like their European counterparts, American meteorologists were at least generally aware of developments elsewhere in the world. Joseph Henry had traveled abroad, but most natural philosophers interested in weather depended on scientific journals and private correspondence. Nineteenth-century science was small and homogeneous, with few specialized institutions. The British Meteorological Society, founded in 1850, was preceded by nearly three decades by the Meteorological Society of London, which was formed in 1823 (Fleming 1990:166–167). But in 1870, when the U.S. government established a weather service, there was no formal meteorological society to serve as advocate and critic.

The Varied Roles of the Weather Map

The weather map's history illustrates a number of significant spurts of scientific and technological progress as well as a diversity of cartographic roles. As discussed in the previous section, the earliest weather maps were tools for scientific discovery and theory development. Only after the electric telegraph enabled the timely assembly of data for many dispersed and distant sites did the weather map become a routine forecasting tool. As it became more reliable and useful, the weather map acquired additional roles, including education, communication, advertising, and even entertainment.

The weather map's educational role reflected the comparatively rudimentary level of meteorological theory in the late nineteenth and early twentieth centuries. With only a few reliable principles, such as the notion of storms as low-pressure cells with counterclockwise circulation, weather forecasting was an intensely empirical enterprise, learned in part through the conscientious study of weather maps. Among a variety of forecasting heuristics in use was Ralph Abercromby's (1887) "seven fundamental shapes of isobars" (Figure 2.4). Abercromby argued that a meteorologist should be able to deduce local weather conditions from these few basic shapes. To educate its staff, the Weather Bureau's Washington office prepared detailed weather maps twice daily, at 8 A.M. and 8 P.M. Printed and bound series of the Washington weather map were distributed to local forecasting offices for study by experienced and neophyte forecasters alike. As one staff meteorologist noted, "the most profound students of atmospheric physics . . . were far from being the best forecasters. . . . In order to excel in the profession one must possess a special faculty for intuitively and quickly weighing the forces indicated on the weather map and calculating the resultant. This special faculty is developed by long and continued study and association with the maps, rather than by a pro-

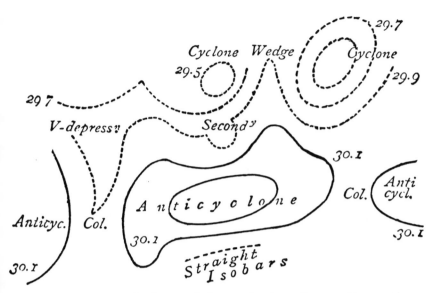

Figure 2.4. The seven fundamental shapes of isobars. Source: Abercromby (1887:25).

found study of atmospheric physics" (Bliss 1917:110). Weather maps were an important part of the Bureau's *Monthly Weather Review,* which included descriptions of noteworthy storms and monthly plots of storm paths, mapped in an effort to discover intelligible patterns.

Because so much of the Weather Bureau's work involved weather maps, the map was an obvious and appropriate symbol of the agency. As one official remarked in 1898, "the daily weather map is the one publication around which the structure of our Bureau is being erected" (Beals and Sims 1899:76). As the number of stations preparing, duplicating, and distributing weather maps rose from 52 in 1891, to 73 in 1893, 84 in 1898, and 112 in 1909, the weather map became an important part of the local forecast. These "station weather maps" served simultaneously as convenient graphic summaries of the nation's weather, displays for educating the public about meteorological principles, and advertisements for the Weather Bureau and its work (Monmonier 1988).

As a result of this public prominence, the graphic-arts quality of the weather map became an important issue at congressional budget hearings when, between 1903 and 1909, the Weather Bureau sought to improve the appearance of its locally published station maps by installing better duplicating equipment and hiring additional experienced printers. Stations serving small and medium-size cities typically reproduced the weather map rather crudely using a stencil-duplicator or mimeograph operated by a clerk or junior forecaster. To demonstrate the need for additional funds, Bureau officials would exhibit several barely legible "duplicator" maps alongside a few more visually satis-factory maps from larger, better-supported stations. Such comparisons were no doubt helpful in winning the support of Congressmen and Senators from areas with aesthetically deficient station maps. In addi-tion to posting them in public places, the Weather Bureau distributed station maps to subscribers by mail, which was delivered more fre-quently and efficiently early in the century.

Weather Bureau officials also referred to the weather map in defend-ing the cost of telegraphy. Each station prepared its own map using data telegraphed from Washington. Because forecasters in smaller cities al-ready received fewer station reports than forecasting offices in large cities, cuts in the telegraphy budget would reduce not only the accuracy of the local station's map but also the reliability of its forecasts.

Pressured by Congress to control costs and eager to reach a wider audience, the Weather Bureau decided in 1910 to encourage local newspapers to print daily weather maps. Figure 2.5 is an example of the small, highly generalized map for which a weather station fur-nished local newspapers either a printing plate or a clear, crisp copy for convenient photographic engraving. The Weather Bureau used the term "commercial weather map" to distinguish its newspaper maps—printed and distributed outside government—from the "station map," a larger, generally more detailed version printed at weather stations. Early response was impressive. By July 1, 1910, 65 newspapers were printing the map in 45 cities, and by 1912, 147 newspapers with a combined circulation of 2.9 million were delivering a daily weather map to readers in 106 cities. The success of the commercial weather map allowed the Bureau to discontinue reproducing the station weather maps in 51 cities. Although a number of these 147 newspapers later dropped the weather map, the commercial weather map circu-lated far more widely than the larger, more costly station weather map, of which 59 stations printed a daily total of only 21 thousand copies (Monmonier 1988).

When meteorological theory made significant gains after World War I, the weather map not only became more reliable but also lost much of value as a symbol of meteorological progress and of the Weather Bureau. Norwegian meteorologist Vilhelm Bjerknes and his colleagues at Bergen developed an improved model of cyclone development and discovered the significant effect of the "polar front" on European and North American weather (Friedman 1989). In the 1920s, Weather Bu-

WEATHER CONDITIONS AT 8 A.M.

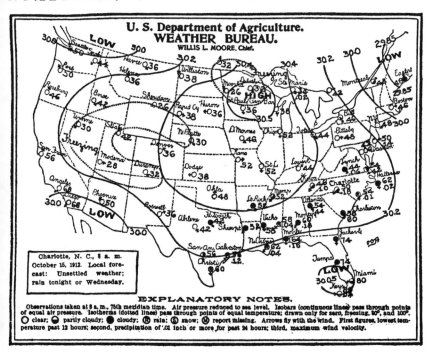

Figure 2.5. An example of the commercial weather map supplied to newspapers by the U.S. Weather Bureau. Source: Heiskell (1912:538).

reau officials testifying at annual budget hearings had little to say about the appearance and distribution of weather maps. Their concern had shifted to the need to sample the upper atmosphere with balloons carrying meteorological instruments and a small radio transmitter. These devices, called *radiosondes*, led to 500-millibar maps and airmass analysis.[3] In the 1930s, a new set of cartographic symbols representing warm, cold, occluded, and stationary fronts enriched the daily weather map. Following World War II further study based on the observations of military pilots led to recognition of the jet stream's influence on weather patterns. Although meteorologists avidly mapped jet streams and other upper atmospheric conditions, the public usually received this information indirectly on a more conventional forecast map showing only surface weather. Until dynamic television weather maps became widespread in the 1980s, few people made the connection between surface weather and the jet stream.

Electronic technology, which made the timely weather map practicable in the 1850s and 1860s, led a century later to significantly new formats. An artificial satellite occupying a geostationary orbit 35,900 kilometers (22,300 miles) above a fixed point on the equator can now monitor cloud patterns and surface temperatures for an entire ocean or continent. Every half hour the satellite's electronic imaging system captures and transmits to earth a revealing cartographic snapshot of storms and atmospheric moisture, and a computer graphics system adds meridians, parallels, continental boundaries, and other reference data and collects these images into the smooth, animated succession of images that television newscasters call the "satellite loop." Equally representative of modern meteorology's approach to mapping is the "computer model," an electronic map of sort that integrates satellite, ground, and radiosonde observations in a massive worldwide database. Some computer models partition the earth's surface into a horizontal grid of cells spaced 100 kilometers (62 miles) apart and sampled vertically at 15 different levels (Burroughs 1991:151–158), but because fine grids and numerous vertical layers slow down computations, many models have a lower resolution and fewer layers. Supercomputers use mathematical simulation models based on hydrodynamic theory to project these weather patterns several days into the future.

The weather map's role has changed markedly since the 1890s. Whereas the forecaster once worked directly with the data and drew a small number of maps, computer models reflecting different assumptions and concepts now generate an abundance of maps, for different times and elevations. Because computer-generated forecast maps often do not agree, the forecaster must decide which assumptions are appropriate and either choose one map or integrate the maps into a hybrid forecast. Vexing at times, this abundance of information has many advantages, especially for consulting meteorologists, who develop specially tailored local forecasts for growers and outdoor recreation facilities. And with the help of weather graphics syndicates employing animation experts as well as experienced forecasters, local television weathercasters can weave these otherwise cartographic products into interesting and informative dynamic narratives.

News publishers have long been enthusiastic users and distributors of weather maps. Although some newspapers that had carried the Weather Bureau's commercial weather map in 1912 dropped it once the novelty had passed, many continued to print the daily weather chart provided by their local forecasting office. In the 1930s, when interest in aviation and progress in air-mass analysis made weather patterns more newsworthy, additional papers started or resumed the daily weather map. In 1935, the Associated Press (AP) inaugurated its

WirePhoto network and offered subscribing newspapers morning and afternoon weather maps redrafted by the AP's Washington office from charts provided by the government. United Press International (UPI), which developed a competing photowire network, also provided timely weather maps for both morning and afternoon newspapers (Monmonier 1989:112–124). After the launch of a series of ESSA weather satellites in 1966, both wire services began transmitting cloud-cover photos from the Weather Bureau. In the 1970s, many television stations invested heavily in weather graphics systems that integrated cloud-cover imagery from weather satellites, richly colored forecast graphics from private electronic meteorological services, and their own radar images of regional precipitation patterns (Henson 1990).

In the later 1970s and early 1980s, the weather map became an essential ingredient in the redesign of the American newspaper. News publishers threatened by increased competition from television for readers' attention sought to package the news more conveniently and attractively. In 1982, many publishers felt threatened by *USA Today*, a new national daily newspaper that used a page-wide, full-color weather map as a key design element, logo, and advertisement. That *USA Today's* weather map lacked both fronts and pressure cells attests to its largely iconic role. Nonetheless, local and metropolitan newspapers responded in a variety of ways, principally by substituting a full-color temperature map for the standard weather map, by dropping the comparatively drab satellite photo, or by adding a regional forecast map with readily interpreted pictorial point symbols showing rainy, snowy, cloudy, or clear conditions (Monmonier and Pipps 1987). A few newspapers, notably the *New York Times*, adopted a highly informative yet less visually prominent component of the *USA Today* weather page, the expository weather map specially designed to explain an important recent or near-future weather event. Ironically, a newspaper's richest, most instructive weather maps often are comparatively small and inconspicuous.

Maps, graphs, tables, short narrative summaries, and other elements of the newspaper "weather package" might satisfy the public's thirst for highly factual, sports-page-like coverage of atmospheric phenomena, but only the dynamic maps of the television weathercast effectively exploit weather's entertainment value. Weather is, after all, a coherent yet constantly changing drama, offering the vicarious violence of a soap opera. Although weather affects us all, usually only in a minor way, severe storms with the destructive potential of atomic bombs and mass murderers can destroy thousands of businesses, homes, and human lives. In late summer and early fall any tropical storm sufficiently powerful to warrant a personal name appears on the cartographic stage like an angry, troubled adolescent: if not directly threatened, viewers are intrigued that Andrew, Bonnie, or Charles might become a history-making meteorological monster. Although most tropical depressions fizzle after racing across the weather map in a couple of anxiety-ridden days, forecasts of the few truly destructive storms that survive fuel public fears. Throughout the year, the weather provides a rich variety of other immediate or vicarious threats: blizzards, spring floods, tornadoes, and devastating droughts. On the television screen, a barrage of satellite maps, forecast maps, radar plots, jet stream diagrams, and other weather graphics make these threats real for millions of viewers thousands of miles away. Cable services such as "the Weather Channel" fulfill the needs of weather enthusiasts, travelers, and others who can't wait for scheduled newscasts. In early 1991, electronic weather maps of southwest Asia heightened the verisimilitude of the Persian Gulf War when weathercasters included a satellite-supported "Gulf forecast" (Clarke 1992).

How the Weather Map Has Changed the World

The idea of mapping weather has changed the world not only by affording an understanding of atmospheric processes but also by making weather predictable, at least in the short run. Although by no means sufficiently timely for forecasting, the first weather maps demonstrated the statistical map's power as a visual scale model convenient for organizing abstract data and revealing geographic patterns. Two-dimensional maps of surface weather allowed Brandes, Loomis, and others to discover fundamental relationships among pressure, wind, and precipitation and to formulate hypotheses about atmospheric circulation and related three-dimensional processes. The weather map has supported a full range of visualization functions not only by encouraging natural philosophers and meteorologists to explore and discover but also by enabling them to predict and confirm. Insights gained by comparing maps of actual and forecast weather led to the development of new instruments, richer theory, and more reliable forecasts.

Electronic technology's role as a persistent catalyst for improving the reliability and foresight of the weather map suggests even more monumental impacts. In the 1980s, meteorologically relevant progress in electronics advanced from data collection to data processing. As the electric telegraph, radiosonde, and geosynchronous weather satellite removed earlier macro-distance impediments to timely maps of current weather, parallel processors, and supercomputers promise to de-molish the "thirty-six hour wall" beyond which weather forecasts are not consistently reliable (Kerr 1990). Moreover, emphasis on computer simulation models should strengthen ties between meteorology and climatology as geographic climatologists and other atmospheric scien-tists concerned with global change attempt to project the present state of the atmosphere not days but decades into the future. By illustrating the likely effects of coal-fired power plants, clear-cutting in the Ama-zon, and other ecologically questionable practices, computer-gener-ated climatic maps might help humankind choose either a benign or a malevolent effect on their descendants' weather. By demonstrating the power of everyday weather and our ability to influence long-term cli-matic change, meteorological cartography is priceless propaganda in the worldwide war against greed and short-sighted complacency (Hall 1992:127–138).

Whether models of global change can win popular support for more ecologically responsible modes of consumption and production may well depend upon how effectively televised weathercasts promote a public understanding of the atmospheric soap opera. As successful teachers are well aware, educating people requires first getting their attention. Entertaining graphic displays can be useful in engaging a class or an audience in material they would otherwise consider dry and complex. For this reason, dynamic weather maps that dramatize and entertain might well prove vital (from *vita*, the Latin for "life") in diverting the world's peoples from an otherwise dismal, disastrous fu-ture of poisoned air, coastal flooding, and devastating storms. By dra-matizing our connectedness with distant places, dynamic weather maps also foster the global awareness needed to resolve growing inter-national environmental conflict.

Notes

1. The coriolis effect is a byproduct of the earth's rotation, which adds an acceleration to any moving object, including winds. This added acceleration reflects the outward force experienced by a person on a carousel or a fly on a turntable. There is no coriolis acceleration at the poles, and at the equator the force is merely outward (that is, upward). But elsewhere on the earth, the cori-olis effect diverts moving objects to the right in the northern hemisphere and to the left in the southern hemisphere. Although imperceptible to a human walking across a floor, the coriolis effect has a substantial influence on the earth's atmosphere, particularly on the pattern of winds around cells of low and high pressure. North of the equator, where diversion is to the right, air

moving outward from a high-pressure cell has a clockwise pattern, whereas air moving inward toward a low-pressure cell has a counterclockwise pattern.

2. Storm movement reflects the earth's rotation and interaction between the lower and upper layers of the atmosphere. Winter storms originate when cold, heavy Canadian air moves southward at the surface into the United States. As the earth rotates from west to east, it drags the atmosphere, including surface air masses, eastward. But high-altitude winds associated with the jet stream move surface storms farther ahead, toward the east. The polar jet stream is the result of warm air that rises from the surface, where the pressure is higher, moves upward in the atmosphere, where pressure is lower, and then northward toward colder air at altitudes of about 40,000 feet. Strong west-to-east wind currents arise as the coriolis effect diverts this northward flow of air to the right. This upper-atmosphere weather then affects surface weather by dragging cold air masses (and their fronts) from west to east.

3. To describe conditions in the upper atmosphere, meteorologists use maps based on high-altitude measurements. One cartographic approach is the 500-millibar map, which uses elevation contours to show heights at which atmospheric pressure equals 500 millibars, the average pressure at roughly 18,000 feet (5,500 meters) above sea level. Pressure falls with increased elevation, and where surface pressure is relatively low, 500-millibar pressures are closer to the ground than where surface pressure is high. A typical upper-air low-pressure cell might be marked by 500-millibar heights of 16,000 feet or less, whereas for an upper-air high-pressure cell the 500-millibar surface might extend about 19,000 feet. Useful for weather forecasting as well as air navigation, 500-millibar maps also show high-altitude temperature and wind direction.

References

Abercromby, R. 1887. *Weather: A Popular Exposition on the Nature of Weather Changes from Day to Day.* New York: D. Appleton.

Beals, E. A., and A. F. Sims. 1899. Relations with the Press, Commercial Bodies, and Scientific Organizations. *U.S. Weather Bureau Bulletin* 24:69–79.

Bliss, G. S. 1917. A History of Weather Records, and the Work of the U.S. Weather Bureau. *Scientific American Supplement* 84(2172):110–111.

Burroughs, W. J. 1991. *Watching the World's Weather.* Cambridge: Cambridge University Press.

Clarke, K. C. 1992. Mapping and Mapping Technologies of the Persian Gulf War. *Cartography and Geographic Information Systems* 19:80–87.

Fleming, J. R. 1990. *Meteorology in America, 1800–1870.* Baltimore: Johns Hopkins University Press.

Friedman, R. M. 1989. *Appropriating the Weather: Vilhelm Bjerknes and the Construction of a Modern Meteorology.* Ithaca: Cornell University Press.

Hall, Stephen S. 1992. *Mapping the Next Millennium.* New York: Random House.

Harrington, M. 1894. History of the Weather Map. *U.S. Weather Bureau Bulletin* 11:327–335.

Heiskell, H. L. 1912. The Commercial Weather Map of the United States Weather Bureau. In *Yearbook of the Department of Agriculture, 1912*, 537–539. Washington, D.C.: U.S. Government Printing Office.

Henson, R. 1990. *Television Weathercasting: A History.* Jefferson, N.C.: Mcfarland.

Kerr, R. A. 1990. Squeezing out Better Weather Forecasts. *Science* 250:30.

Loomis, E. 1846. On Two Storms which Were Experienced Throughout the United States, in the Month of February, 1842. *Transactions of the American Philosophical Society* 9:161–184.

Marriott, W. 1901. An Account of the Bequest of George James Symons, F.R.S., to the Royal Meteorological Society. *Quarterly Journal of the Royal Meteorological Society* 27:241–260, esp. 258–259.

———. 1903. The Earliest Telegraphic Daily Meteorological Reports and Weather Maps. *Quarterly Journal of the Royal Meteorological Society* 29:123–131.

Miller, E. R. 1933. American Pioneers in Meteorology. *Monthly Weather Review* 61:189–193.

Monmonier, M. 1988. Telegraphy, Iconography, and the Weather Map: Cartographic Weather Reports by the United States Weather Bureau, 1870–1935. *Imago Mundi* 40:15–31.

———. 1989. *Maps with the News: The Development of American Journalistic Cartography.* Chicago: University of Chicago Press.

Monmonier, M., and V. Pipps. 1987. Weather Maps and Newspaper Design: Response to USA Today? *Newspaper Research Journal* 8(4):31–42.

Robinson, A. H. 1982. *Early Thematic Mapping in the History of Cartography.* Chicago: University of Chicago Press.

Scott, R. H. 1875. Weather Charts in Newspapers. *Journal of the Society of Arts* 23:776–782.

Shaw, N. 1926. *Manual of Meteorology*, vol. 1, *Meteorology in History.* Cambridge: Cambridge University Press.

True, W. P. 1929. The Beginnings of the Weather Bureau. Chap. 18 in *The Smithsonian Institution*, 299–310. Washington, D.C.: Smithsonian Institution Series.

3

Geographic
Information Systems

Michael F. Goodchild

In Chapter 1 Anne Godlewska explored the idea of the map and showed how the development of printed representations of the world changed our view of our surroundings. The role of maps in structuring our thinking about the world is often subtle but nevertheless real, and recent work by cartographers has helped to draw attention to the power of maps to influence the relationship between people and their environment (Wood 1992). Although we often think of maps as pictorial, and therefore neutral, representations of the world, the choice of features, the ways in which they are differentiated, and the way in which the curved surface of the earth is distorted to fit onto a flat sheet of paper are all representational choices that can powerfully influence our understanding of the world. Even the custom of placing north at the top of a map has been seen as sending a subtle cultural message.

Over the past three decades, as part of a more general trend toward the use of digital technology for information handling, a significant change has occurred in the nature of geographic information and its role in society. Unlike numbers and text, maps and images have created major problems for digital storage and processing because of their relative complexity and high density of information. The so-called "geographic information technologies" that have appeared over the past twenty years and grown to form a major new area of application in electronic data processing are the result of significant research developments in hardware and software. The new technologies are likely to be every bit as influential as the map has been on our thinking about the world. They broaden our perspective by offering new capabilities that are less constrained, but at the same time they impose new filters that may be just as subtle as the ones associated with paper maps.

Limitations of Traditional Mapmaking

The earliest symbolic representations of the world around us made use of a variety of materials, from scratches on rocks to stick models used to navigate the Pacific (Harley and Woodward 1987). But the emergence of paper and printing created a medium that came to dominate mapmaking, at least until the development of aerial photography and digital imagery in the twentieth century.

With simple tools, it is possible to make representations on paper of a wide range of geographic phenomena and features and thus to communicate knowledge of them to the person reading the map. It is possible to print text using movable type or templates, to draw lines of constant width, and to fill areas with uniform color. Additional techniques allow areas to be filled with uniform, pre-printed patterns, such as dots or cross-hatching. But despite their apparent flexibility, these techniques are in reality quite limiting (Goodchild 1988). It is difficult, for example, to vary line width continuously in order to represent changes in the width of a road or river or to indicate varying uncertainty about the position of a boundary line or contour. It is hard to fill with continuously changing color or pattern in order to represent gradients in the densities of species at the edge of a forest or in rainfall in the lee of a mountain range. It is also difficult for the map reader to interpret continuous variation correctly. As a result, map representations tend to emphasize the abrupt or crisp aspects of geographic variation and to suppress smooth change or fuzziness.

Nowhere is this sense of technical limitation stronger than in the conventions devised to depict topographic elevation. The familiar isoline or contour technique has been standard for many decades, but its limitations are clear in the need for instruction in map reading in schools and in the use of supplementary techniques such as color or hachuring to help the map reader interpret the pattern of contours correctly. Isolines evolved not only because of their efficiency in communicating information about topographic elevation, but also because they were one of the few methods available within the constraints of cartographic technique; they could be drawn with a pen. They give no information at all about the shape of the surface between contoured elevations or about the cartographer's uncertainty regarding elevation

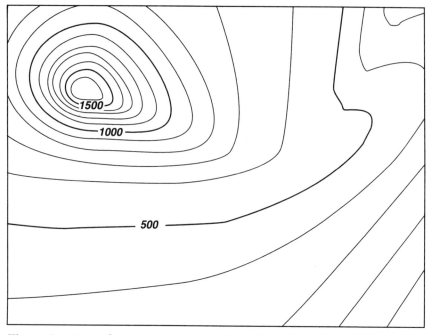

Figure 3.1. Part of a contour representation of a surface. Mapmaking technique requires that contours be of uniform width, and hence cannot represent either varying uncertainty in contour position or the form of the surface between contours.

measurements and the consequent impact of that uncertainty on contour positions (Figure 3.1).

If the objective of the map is to communicate the knowledge of the mapmaker to the map reader, then one can reasonably ask questions about the efficiency of the communication. In the case of contours, the information available to a mapmaker following conventional topographic mapping practice includes direct measurements of elevation at selected points (spot heights), plus the ability to estimate elevation photogrammetrically at any point. The map reader is provided with estimates of elevation of known accuracy at points lying on contours, plus the ability to estimate elevation at other points with much less accuracy. For much of the map, elevation could be anywhere between the elevations of the adjacent contours, so the quality of estimates depends directly on the contour interval.

In short, because only a small fraction of the information available to the mapmaker is actually communicated, the efficiency of map communication can be disappointingly low. Similar points can be made about the efficiency of soil maps in communicating the knowledge of field surveys by soil scientists or about choropleth maps as communicators of knowledge about population density. In all of these examples, the information available to the map reader is only a small fraction of that available to the mapmaker, at least in part because of the constraints imposed by the technical limitations of cartographic technique. Because the real world is a complex and often frightening place, however, a map that simplifies the world and communicates only a small fraction of its real complexity may actually provide the map reader with subtle reassurance.

Mark and Frank (1991) and others have recently introduced principles of cognitive science and linguistics into discussions of the nature of geographic information and mapmaking. These scholars suggest that our ability to learn and reason about geographic information is limited by the structures inherent in language, structures that are strongly biased toward a view of the world in terms of discrete, crisp objects and sharp change. Such patterns are the geographic expression of a more general propensity to use discrete categories and to attach labels. The English language, for example, is rich in terms describing spatial relationships between objects (terms such as "within," "over," "across," "outside"), but fuzziness, uncertainty, and continuous change tend to be avoided in everyday human discourse and are less well served by English vocabulary.

These arguments suggest that the use of contours to depict elevation, or of crisply defined areas to depict soil variation, reflects human cognitive preference as much as the limitations of cartographic technique. In this view, the filtering imposed by the communication channel is considered not an unfortunate loss of information, but rather a desirable stage of generalization, a way of improving the usefulness of cartographic information to the map reader. A map that showed the world as it really is, in all its complexity, fuzziness, and spatially continuous change, would be less useful than the conventional product because its contents would fail to fit with our cognitive processes. In short, we find it difficult to learn about the world or to describe it to others in terms of grayness or fuzziness; we also have difficulty reasoning with such information.

This tension between an enlightened, scientific view of maps as imperfect communication channels and a cognitive view based on an analysis of the limitations of human reasoning is central to debates over the role of new technologies. On the one hand, computer technology has the potential to remove the constraints of cartographic technique, opening the prospect of a brave new world in which the scientific knowledge of the soil scientist, for example, will be fully available to others through novel forms of representation. On the other hand, computer technologies designed to reflect our limited abilities for spatial reasoning can make maps easier to read, and can move map design into closer compatibility with human intuition. Clearly there is room for both viewpoints, depending on what one is willing to assume about the context of application. In the remainder of this chapter I look more specifically at some of the new ideas that have emerged with the development of computer-based tools for mapping and geographic information handling and consider their potential impact on our understanding of the world.

The New Geographic Information Technologies

Geographic Information Systems

The term geographic information system was coined in the 1960s. In Canada, Roger Tomlinson was leading the effort to build a computer-based system to handle the mass of geographic data created by the Canada Land Inventory (Tomlinson, Calkins, and Marble, 1976). A land inventory must focus on measures of land area, such as the acreage available for growing arable crops, measures that are difficult to obtain from maps by the conventional methods of dots and counting, or tracing around areas using a mechanical device called a planimeter, both of which provide estimates that are crude at best. To make matters worse, land-use planning requires the accurate estimation of areas of land having multiple characteristics, which would have to be obtained from the various maps in the Canada Land Inventory through the labor-intensive preparation of transparent overlays. In planning forest use, for example, one would want to know the capability of land not only for forestry, but also for recreation and agriculture. A digital computer offered a potentially cost-effective and much more accurate alternative. The term geographic information system was coined to describe the system's emphasis on the geographic dimension of data, and its ability to store, retrieve, and analyze a wide variety of data types through simple manipulations (see Maguire 1991, for a detailed discussion of definitions of GIS).

From these early beginnings, geographic information systems have grown to become a significant area of electronic data processing. Recent figures place the total annual U.S. expenditure on GIS software at around $450 million (Daratech 1994), and expenditures on associated hardware, data collection, maintenance, and analysis are far higher. Many of the research breakthroughs associated with GIS occurred during the development of the early systems, particularly Tomlinson's Canada Geographic Information System (Tomlinson, Calkins, and Marble, 1976), and the products of the Harvard Laboratory for Computer Graphics and Spatial Analysis, which flourished in Cambridge during the 1970s. Commercial products incorporating these developments began to appear in the late 1970s, and by the late 1980s GIS research had spawned a series of conferences, magazines, societies, and university and college programs. The field continues to expand, with software sales rising by 20 percent or more annually, with no end in sight. GIS has been described as "simultaneously the telescope, the microscope, the computer, and the xerox machine of regional analysis and synthesis" (Abler 1988:137) and as "the biggest step forward in the handling of geographic information since the invention of the map" (Department of the Environment 1987:8).

A modern GIS such as ARC/INFO, a product of Environmental Systems Research Institute of Redlands, California, or Geographical Resources Analysis Support (GRASS), developed by the U.S. Army Construction Engineering Research Laboratory of Champaign-Urbana, Illinois, has facilities to store digital representations of a wide range of geographic features and types of geographic variation. Besides geometric representations, a GIS can also store a range of attributes of each feature (in other words information that is known about each feature and serves to distinguish it from others) and also a range of relationships between features, such as connectivity, adjacency, or proximity. Capabilities exist to transform data from one projection to another, to input data from a variety of different systems and devices, and to display data in cartographic form. In addition, the true power of GIS lies in its ability to analyze data by measuring areas, overlaying different data sets, or carrying out a range of standard types of spatial analysis or modeling. A GIS is a tool for conducting geographical analysis, just as a statistical package is a tool for conducting statistical analysis; both provide researchers with the means to implement a body of well-defined analytical methods.[1]

At its simplest, a GIS database provides a digital representation of the contents of one or more maps. Because features shown on maps are invariably crisp, with sharp, well-defined boundaries, creating digital representations has been possible by using combinations of points, lines, and areas. Points are represented by pairs of coordinates, lines by ordered sequences of points assumed to be connected by straight lines, and areas by ordered sequences of points forming the vertices of polygons. The term polyline is often used to describe this common approach to line representation, through the obvious analogy to polygon. Thus GIS data structures not only preserve the essential crispness of mapped geographic features but abstract them further by insisting on straight line segments where reality may exhibit smooth curves. Although polylines and polygons are appropriate ways of representing city streets or land parcels, in the case of a river, for example, the conventional GIS representation of its plan form shows little understanding of the effects of hydrologic processes, and in this sense one might argue that GIS data structures are less effective representations than maps, because a cartographer is able to draw a smooth curve. Similarly, in the case of land cover maps, GIS designers have typically preserved the sharp transitions between types that are conventional on land cover maps (for example, between forest and open grassland, or between conifer forest and deciduous forest), rather than search for representations that can portray spatially continuous transitions.

Remote Sensing

The earliest aerial photographs date from the last century, but the advent of earth-orbiting satellites in the 1960s opened an entirely new range of technological possibilities. Digital satellite images of earth made it possible to create databases of comprehensive global coverage, without concern for political boundaries. The military applications of remote sensing fueled much of its development, but civilian applications followed, enabling a new perspective on mapping.

Unlike maps, remotely sensed images are composed of rectangular picture elements, or pixels, each carrying a signal representing the radiation from that area of the earth's surface in a particular spectral range. Instruments such as the Thematic Mapper (TM), mounted on the Landsat satellite, return images with a high level of spatial resolution (approximately 30 meters in the case of TM) and overfly any area at regular intervals (nineteen days in the case of Landsat).

Every image from space is constructed from thousands of measurements of radiation intensity. Color and texture changes continuously, and it is not until the image is classified by assigning each pixel to one of a limited number of classes that it begins to resemble a map. Further processing is necessary, such as the filtering of "salt and pepper," or instances of isolated pixels surrounded by pixels of another class, before the world as seen from space begins to resemble the more familiar map representation.

When not inhibited by cloud cover, remote sensing is capable of identifying a wide range of geographic variables with varying degrees of certainty. In recent years data derived from remote sensing have been used as input to analyses and modeling carried out in GIS, where they can be merged with information on features and characteristics not identifiable from space, such as demographic and economic variables, political boundaries, or land ownership. The problems of integrating GIS and remote sensing technologies (and reconciling inherently different perspectives on the world) include accuracy, incompatibility of definitions, and institutional issues (Star 1991).

Global Positioning Systems

In the past five years, remote sensing and GIS have been joined by a third digital geographic technology, now known as the global positioning system (GPS) (Leick 1990). Like remote sensing, its development was funded in large part by the military (Smith 1992 discusses the military role in the development of GIS). The system consists of a constellation of earth-orbiting satellites, each emitting precisely timed signals, which can be used by a ground receiver to compute position on the earth's surface. In military use, a simple GPS receiver can determine position to 32 meters with 90 percent reliability, but for civilian use the signals are corrupted and the 90 percent reliability distance is approximately 100 meters. Receiver technology has now advanced to the point where hand-held receivers the size of a pocket calculator are available for a few hundred dollars, and GPS boards are available for installation in personal computers.

Various techniques have been devised to improve the accuracy of GPS and to circumvent the corruption of the signal. Differential GPS requires use of two receivers, one at a fixed and known location and one roving, communicating by a radio frequency signal. With such systems, positional accuracies of 1 meter or better are achievable, and even higher accuracies are possible if one is willing to collect data at the unknown location for an extended time.

GPS is having a revolutionary effect on the profession of surveying. Traditionally, positions are established by taking measurements with respect to monuments of known position, and each nation maintains a hierarchy of such monuments, including a small number the positions of which are known with great accuracy. GPS allows position on the earth's surface to be determined directly, and its use is leading to the identification of numerous errors in previously determined positions, particularly in remote areas. For scientists working in comparatively featureless areas such as Antarctica, GPS is providing the first reliable way of fixing observations in space.

Like remote sensing, GPS has been integrated with GIS as a source of measured coordinates. Again like remote sensing, its use raises numerous issues of accuracy, professional practice, and terminology.

Taken together, the three new digital geographic information technologies—GIS, remote sensing, and GPS—amount to a massive change in the availability, reliability, and value of geographic information to society. This change has enormous implications for how we represent and understand the world.

New Ideas of Geographic Representation

All three of the geographic information technologies have opened new possibilities for geographic representation. As we have already seen, the representations available to the traditional cartographer, working with pen and ink, are limited, although one can argue that the effects of these constraints are, nevertheless, consistent with patterns of human reasoning. In the digital environment no such constraints exist, and representations are limited only by the imaginations of their creators. In this section, I review some of the key developments of the past three decades.

Features and Attributes

Maps can be designed to show the locations of features, and a number of techniques have been devised for differentiating features or displaying their attributes. One can replace points by symbols denoting, for example, towers or fire hydrants; one can color line symbols or code their characteristics through patterns of dots and dashes to distinguish roads from railroads, for example; and one can fill area symbols with cross hatching or color to denote land cover class or political status. But in each case there are strict limits to the number of attributes that can be displayed simultaneously for a given object. By combining color fill, cross hatching, and labels, it may be possible to show as many as five independent attributes of area objects, but this is, of course, far less than the number of attributes that may be available for certain types of areas. For example, hundreds of attributes may be available for each reporting zone used by the census, including average family income and incidence of a certain disease, but only a very small number can be shown simultaneously on the same map. The clever facial expressions shown in Figure 3.2 are perhaps the most ingenious method for displaying many attributes simultaneously within the constraints of mapping technology.

GIS offers a much richer environment. One commonly used technique is to allow the user to point to a specific feature on the screen, such as a census reporting zone, with a mouse or other pointing device. The database is then searched for attributes of the feature (for

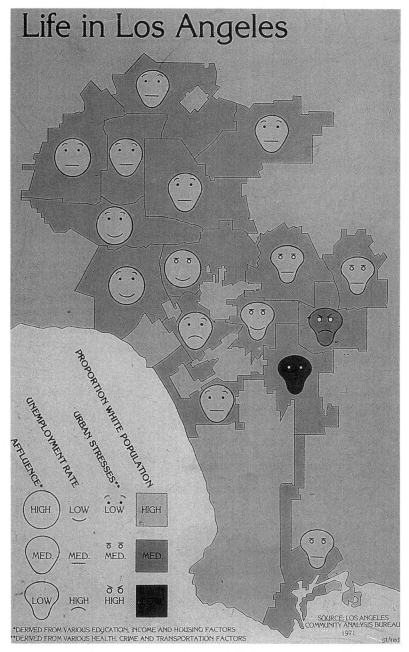

Figure 3.2. Facial expressions send many different messages. In this technique, developed by Chernoff (1973), caricatures of faces are used to display many more attributes than can normally be shown simultaneously on a map. Source: Muehrcke and Muehrcke (1992).

example, the age characteristics of the population over sixty-five in a county, or information about the owner of a parcel of land), and these are displayed in an unused part of the screen. In contrast to a printed map, this gives the display a dynamic aspect, allowing the user to change the visible information on demand. As a result, a far greater range of information can be made available than would be possible in the world of conventional mapping. Through this technique, the user can feel empowered and less dependent on the mapmaker's agenda. If taken too far, however, this new access to geographic detail can cross an invisible line to become an invasion of privacy. For example, it is possible to take data sets already available to the public, such as telephone directories and street maps, and create simple GIS applications that provide names and telephone numbers of house occupants from a map in response to a click of a mouse.

Hierarchical Access

Display of attributes on demand is one way of overcoming the space limitations of two-dimensional maps. Geographic reality has the interesting property that the more closely one looks at the surface of the earth, the more detail one sees (some teachers of geography have even been known to impress this fact upon their students by requiring them to map the intricacies of a 0.25 square meter plot of campus lawn). It is sometimes possible to predict this unraveling of detail, at least for certain cases of natural geographic objects such as shorelines, with some degree of accuracy. This property is so inconsistent with traditional thinking in geometry that it led Benoit Mandelbrot (1982) to propose a general theory of what he called fractals, or objects that display this kind of behavior. Of course map representations do not have this property; the detail that can be included for a given map scale is always limited, because it is difficult to represent objects if they would be smaller than approximately 0.5 millimeters across on the map, however large they might be on the surface of the earth. The Hitachi Corporation has demonstrated the potential of current etching technology by creating an image of the earth on a 5-centimeter wafer of silicon, with detail on the individual streets of London (at this scale a street is about ten atoms across), but such technical marvels remain quite useless.

Unlike a map, a spatial database has no scale, there being no distances inside a digital store to compare to distances on the earth's surface. Thus many GIS displays make use of a magnifying glass metaphor, by providing the user viewing a display at a given scale with an icon that allows any area of the map to be selectively enlarged.

Already-visible features can be shown in greater detail with more attributes, and new features that were invisible at the original scale of the display can be added. Of course this process is limited by the level of detail available in the database, but it provides a powerful new way of looking at geographic variation.

Mapmakers have met the need for views of the world at different scales by publishing multiple map series. Thus topographic maps of the United States are available at 1:24,000, 1:100,000 (for the conterminous United States), 1:250,000, and 1:2,000,000.[2] The definitions used and sets of features shown are different for each series, and the processes of mapmaking are largely independent of each other. As a result, many multi-scale GIS databases preserve this independence, providing no logical linkages among maps of the same area at different scales. For example, although digital representations of the railroad network are available at 1:100,000 and 1:2,000,000, the contents are so different that it is sometimes hard to tell that one is looking at the same area. Although much effort has gone into developing methods for automatically creating small-scale maps from large-scale maps and into "fusing" the information contained in both, it has proven exceedingly difficult to capture the complex and domain-specific processes of generalization used by cartographers into programs for digital computers (Butterfield and McMaster 1991).

Instead, current GISs contain an increasing number of concepts of hierarchy, or explicit relationships between representations at different scales. The quadtree (Samet 1990a 1990b), for example, provides a method of representing variation at multiple scales that has no precedent or analog in traditional mapping. Figure 3.3 illustrates the construction of a simple quadtree representation of the spatial variation of a variable such as land cover class. If the entire area to be mapped contains more than one class, it is subdivided into four equal quadrants. Each is then examined to see if it is homogeneous with respect to land cover class, and if not, it is subdivided again. The process continues until the map consists of a series of quadrants of varying size, each of which is either homogeneous with respect to land cover class or is of the smallest size available given the spatial resolution of the data. Figure 3.3 shows the quadtree representation as a map and as a tree structure.

The quadtree is widely used as an economical representation that adapts the size of its basic geographic units to the degree of local variability in the mapped variable. It is used in the SPANS GIS (Tydac Technologies, Ottawa, Canada), in Samet's QUILT, and as a method of compressing images so that they can be transmitted over networks in

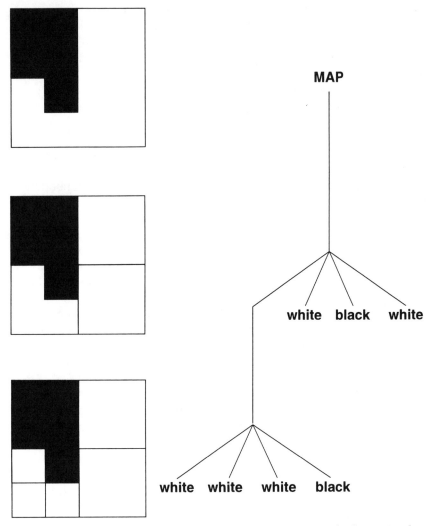

Figure 3.3. An example of a quadtree representation, as applied to a simple two-class (black and white) map. The map as a whole forms the root (top) of the tree and is first divided into four equal quadrants. If any quadrant includes more than one class, it is divided again into four, and the process continues until every subdivision contains only one class. Complex patterns require more levels in the tree than simple patterns, and the number of levels of the tree may also be determined by the basic cell size or resolution of the data.

less time. The quadtree concept has been generalized to the curved surface of the earth (Goodchild and Yang 1992), and similar tree-like structures have been devised for complex linear objects and for indexing discrete features (van Oosterom 1993). Recently there has been much interest in the concept of wavelets (Chui 1992), which can be seen as a generalization of the quadtree in which variation within each quadrant is described by a simple wave function. In all of these cases, a hierarchical tree is used to create a representation with information at all levels of spatial resolution, a concept that is radically different from the uniform resolution of traditional maps or earth images.

Hierarchical approaches to the structuring of geographic data are radically different from the pictorial perspectives provided by maps and images of uniform scale. They give us greater opportunity to explore interesting areas in more detail, or to step back and place areas in their regional context. Hierarchical structures give us access to the important local details that may not be visible at a given scale, and they emphasize the nature of generalized maps as approximations to the truth.

Scan Orders

Remotely sensed images convey information through assemblages of picture elements or pixels, and it is conventional to order such images by row from the top left corner, by analogy to the convention of many, but not all, written languages. This approach is echoed in the data structures of so-called raster GISs, which similarly represent spatial variation by dividing space into rectangular or square cells. In contrast, a vector GIS represents variation by describing the geometric form and attributes of discrete geographic features, either points, lines, or areas. For example, three approaches are commonly used to represent topographic elevation in GIS. The Digital Elevation Model (DEM) uses a square array or raster of point elevations. Digital Line Graph (DLG) representations, often obtained from topographic maps, capture topography in the vector format of digitized contour lines. Finally, Triangulated Irregular Networks (TINs), a second vector option, cover the surface with a mesh of irregular triangles, and assume that the surface varies linearly within each triangle.

Because vector databases prescribe no specific order for the features they contain, it is possible to think of them as scanning space in random order, by contrast to the systematic row-by-row scan of raster databases. Perhaps there is an analogy between vector databases and the action of the eye in scanning a work of art, as it views the image in a sequence that is subtly controlled by the artist. Conversely, a row-by-row scan is analogous to the way images are transmitted by television, or constructed by a computer display.

Early in the development of GIS, it became clear that one might achieve certain objectives by scanning space in orders that had no analogy in written text. In the Canada Geographic Information System (CGIS), data derived from each map sheet were stored on tape. The time required to find the data corresponding to a given map sheet could be substantial if it required winding through long lengths of tape or waiting for new tapes to be mounted. The system's designers argued that one could anticipate a certain pattern of requests in many applications, specifically that the next map needed was likely to be spatially adjacent to the one currently being processed. This principle seemed likely to be valid over a wide range of processes, from basic data editing to query and display. It followed that the optimum storage sequence would be one in which map sheets that were adjacent in space were most likely to be adjacent in storage.

Of course it is impossible to find a linear ordering of a two dimensional space that preserves spatial adjacencies, because a map sheet can be adjacent to at most two map sheets in a linear order but is mostly adjacent to four in space (defining adjacency in the sense of shared edges). The order devised by Guy Morton (1966) and illustrated in Figure 3.4 was selected as the best possible option and implemented in CGIS. It proves to be intimately related to the quadtree (Samet 1990a, 1990b), and Mark (1990) has compared it systematically to other scan orders using a variety of statistics. By coincidence, Morton is also the name of the southwesternmost county in Kansas.

The human eye is a remarkably effective processor of two-dimensional information, but other senses and methods of communication are no better than linear ones. Our administrative tools, such as filing cabinets, lists, and tables, are also largely linear. The development of the Morton scan order is just one example of how GIS research is stimulating new thinking about the relationship between linear and spatial approaches to the organization of information, and the importance of efficient methods for conversion between them.

Time

Because a map is static, cartographic representations clearly favor those aspects of geographic variation that are similarly static or change only slowly through time. Estes and Mooneyhan (1994) have recently commented on the age of much U.S. topographic mapping, which includes many sheets that have not been updated for more than fifteen years. Certain dimensions of topographic maps clearly deteriorate more rap-

Figure 3.4. The Morton order, devised to ensure that areas close together in space are also stored close to each other in the database.

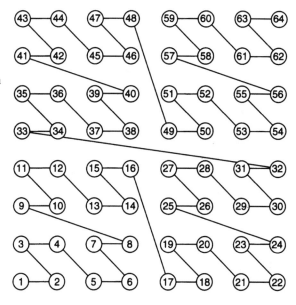

idly than others: although the physical form of the surface may remain more or less constant, cultural features are much more transitory.

Recently, there has been much interest in adding the temporal element to geographic databases. Remotely sensed images are snapshots taken at regular and frequent intervals, and sophisticated methods have been developed in GIS for detecting change between images, while allowing for the effects of seasons and sun angle. It seems that many of the functions currently performed by topographic maps could equally well be performed by suitably corrected air photos, and in recent years the U.S. Geological Survey has developed the necessary methods for producing digital orthophoto quadrangles (DOQs), which are air photos corrected to vertical perspective, with a pixel size of 1 meter. Because DOQs can be produced for less than one-tenth the cost of a standard topographic quadrangle, frequent updates in areas of rapid geographic change are cost effective. Relative to standard topographic maps they lack only the identification and interpretation of features. With increasing use of satellite images and DOQs, the orthographic but uninterpreted view from space is likely to become the basis of much of our perspective on geography in the future.

Langran (1992) discusses alternative approaches to the representation of temporal change. In the simplest case, time provides a series of snapshots, or layers of information in the GIS database, with no explicit logical relationships between the contents of adjacent time slices. In the second case, which is equally easily handled within the design of current GIS, features persist in fixed locations through time, but their attributes change in longitudinal series. Much census data follows this model, in that the decennial census provides a regular update on the contents of a series of fixed reporting zones. The third case is characterized by features with persistent identity, but with positions and attributes that change through time. Change of position may be described by movement parameters of speed and direction or by knowledge of the feature's position at fixed points in time. This model has been used to represent space-time behavior of individuals (Goodchild, Klinkenberg, and Janelle 1993) or populations of animals in zoological applications of GIS, but the functions and data model concepts necessary to support it fully do not yet exist.

In the most problematic case, features have identity at points in time, but identities fail to persist as features move, dissolve, coalesce, and split. Although it may be possible to count the number of clouds on a single image, it is clearly not possible to model clouds as discrete, well-defined features through time. Thus the only available approach to handling this type of data is to fall back on the simple time slice model, the first case discussed earlier. Households show analogous characteristics, as they emerge through marriage and other household formation processes, persist through time, and then dissolve through death, divorce, or other forms of fragmentation.

Although none of these approaches is in any way original as a model of time-dependent information, the treatment of time within GIS represents something of a conceptual breakthrough, given the static nature of traditional maps and the prevalence of the map metaphor within early GIS architectures. It remains to be seen whether the development of better methods for representing time in spatial databases will lead to a rethinking of associated methods of data collection. With new storage and processing technology, is it still appropriate to collect census data every ten years, or should we move to some more continuous scheme, perhaps based on a smaller sample?

Exploratory Spatial Data Analysis

By the late 1970s, the computer had become a widely used tool for statistical analysis, allowing the analyst to avoid the tedious calculations that had characterized early applications of such techniques as factor analysis. As the tool stimulated new ways of thinking, new methods of analysis were developed that capitalized on its power.

Nowhere was this more apparent than in the development of exploratory data analysis (EDA), a set of techniques pioneered by Tukey (1977) and others. Whereas conventional statistical analysis stressed the formulation of hypotheses, followed by formal testing, EDA offered tools for exploration of data as part of the process of hypothesis generation, together with new methods for visualizing samples from interesting perspectives. Today, the EDA paradigm is widely accepted, along with related approaches that engage the computational power of the digital environment.

The advent of GIS has raised the prospect of similar opportunities in spatial data analysis. Because maps are tedious and costly to produce, much traditional analysis of spatial data has tended to ignore the spatial component, treating reporting zones as if they were independent samples from a statistical population and adopting statistical techniques based on the assumption of independence.[3] The consequences have been documented repeatedly (see, for example, Openshaw 1983; Haining 1990; Anselin 1989).

Exploratory spatial data analysis (ESDA) applies the EDA approach to spatial analysis (see, for example, MacDougall 1992). These exploratory methods include tools to display multiple perspectives on data in the form of maps, time series, tables, and charts; active visual linkages between perspectives, such that pointing to a data element on a scatterplot will simultaneously highlight the same element on a map; hierarchical linkages between and among data that allow analysis at multiple scales; and animation. Fotheringham and Rogerson (1994) provide a review of recent GIS-based efforts to find novel approaches to spatial data analysis.

For decades, geographers have argued over whether the purpose of their research should be to search for general laws that apply everywhere on the earth's surface, or for understanding of the particular characteristics of places. The former paradigm is strongly associated with the scientific, quantitative philosophy of geography that flourished in the 1960s, while the latter is linked to an older focus on regional exploration and description and is currently echoed in postmodern thought. If laws are the same everywhere, then selecting a study area for research is analogous to selecting a sample and can be assumed to have minimal effect on the outcome. Most of the tools of statistical research that were widely adopted by geographers during the quantitative revolution of the late 1950s and 1960s were aspatial, ignoring the geographic locations of the cases being analyzed. But GIS tools capture geographic locations, giving the user the choice between search for general laws and exploration of geographic uniqueness. They allow us to look for exception to theories or to ask, for example, whether theories that seem to work for suburban areas also work for rural areas.

Uncertainty

Although techniques for expressing ignorance and uncertainty were commonly used in early maps (for example, the blank areas or "terra incognita" in the interior of Africa), modern mapping preserves few if any ways of communicating lack of knowledge or of warning the map reader of possible inaccuracy. The border between the Yemen and Saudi Arabia is undefined and is commonly shown as dashed; intermittent streams are marked with the same dashed symbol to indicate uncertainty over the presence of water. Dashes are also commonly used to indicate planned features or features under construction. With these few exceptions, maps continue to reassure us that the information they present is certain.

In reality, the quality of the information shown on maps may be uneven and uncertain. Soil maps, for example, show areas of uniform class separated by thin lines representing sharp transition, although it is clear that transitions are often far from sharp, and areas are often far from homogeneous. Such information may be available to the informed reader in the legend, information that often accompanies the map, or in separately published data-quality statements. But when the map is digitized into a GIS database, such qualifications are unlikely to be readily available to the data's eventual user.

To be fair, many maps were never intended to be repositories of scientifically objective measurements. Contours drawn on topographic maps are intended to convey an impression of the form of the surface, not to capture precise elevations. But the unwary user of a GIS database may well believe that the response received to a query about elevation at a point is indeed a scientific measurement, with an accuracy implied by the degree of detail reported by the system. In such strange ways does the GIS confer accuracy and authority on the data that it contains.

Recent thinking on the subject of data quality in GIS has converged on visualization as the appropriate method for communicating a meaningful sense of qualification to the user. Colors can be diluted toward gray to indicate lack of confidence in attributes. Contour lines in a computer display can be broadened or blurred to reflect the positional uncertainty that is a logical consequence of uncertainty in knowledge of elevation. Boundaries between classes on soil or land cover maps can be blurred to indicate transition zones, and randomly shaped inclusions can be placed within patches to indicate knowledge of heterogeneity. The important message that such random inclusions are not the truth, but one possible version of the truth, can be conveyed by displaying a series of possible maps on the screen, or by animation. In effect, research is reverting to the days when mapmakers routinely portrayed uncertainty by leaving areas blank, but a GIS can convey uncertainty in a more rigorous and useful fashion.

Conclusion

New tools have often provoked new thoughts, and recent digital geographic information technologies are no exception. Traditional technologies used to make maps impose constraints on how their users see the world; those constraints may coincide with an innate human desire for a simpler, more ordered environment. The newer digital technologies are not so constrained, and so open up a rich new set of options for representing the world. As a result, new kinds of thinking have emerged through the interplay among geographic information technologies, their developers, and their users.

Even the most elegant cartographic techniques allow only a small number of attributes to be displayed for each feature shown on a map, and map scale imposes limits on the sizes of features that can be shown. Because a digital database has no scale, it imposes no limits on the density of information. A digital store does not require that large amounts of space be devoted to empty desert, nor does it require simplifying the complexity of dense urban areas, or limiting the amount of information that can be associated with a single feature. Maps tend to emphasize the horizontal relationships among features of comparable extent and to equate extent with importance. A geographical database can give access to the relationships between large and small features and can draw attention to the importance of the small and particular as well as the large and general.

Maps are inherently two-dimensional and static, whereas digital technologies can capture the three-dimensional and time-dependent aspects of geographic variation. A two-dimensional map cannot show vertical differentiation in the atmosphere, or the weather effects that result, or the behavior of a three-dimensional plume of polluting hydrocarbons underneath a leaking underground fuel tank. The ability to handle three-dimensional data in digital geographic databases will encourage us to investigate the subsurface environment and to think about the atmosphere as a three-dimensional system. Adding the temporal dimension to databases will encourage us to think in terms of continuous change in the geographic distribution of human populations, rather than the periodic time slices provided by the traditional census.

GIS and the new techniques of exploratory spatial data analysis allow us to examine information in its geographical context and to ask how conditions change from one area to another. New methods of spatial analysis, such as the technique developed by Openshaw and his colleagues to search for clusters of disease, emphasize the importance of exploration, and the need to combine the intuitive power of the human eye to search for patterns with the rigor of statistical analysis (Openshaw et al. 1987).

Finally, the simple action of analyzing large amounts of geographic data of comparatively low accuracy in very precise computing machines has forced us to think hard about the role of maps as repositories of scientific measurements. We clearly do not know the world as accurately as we thought we did. This realization is forcing us to rediscover earlier methods of communicating uncertainty and to invent new ones in order to caution the users of geographic data technologies against ascribing excessive accuracy to their results. Once again, the world is proving to be not as simple as our maps had led us to believe. The impact on our map-based systems of land ownership, regulation, and taxation remains to be seen.

The ability to store the contents of maps in computers will continue to suggest new ways of thinking about the world for years to come. In this sense GIS has only just begun to influence the myriad ways in which society makes use of geographic information. We have yet to see how the ability to display digital road maps in automobiles will influence road map design or the way people explore and navigate in strange environments. Will the widespread adoption of multimedia computers in American homes lead to a richer knowledge of the rest of the world or one that is controlled even more than at present by simple stereotypes? The ideas that have emerged from GIS are currently limited to the research community and the agencies that have had the resources to acquire the technology. But we are now entering an era of mass marketing of GIS, using simpler tools and aiming at broader applications. The impacts of these tools will make fascinating topics of study for geographers (Pickles 1995).

The emphasis in this chapter has almost inevitably been on the novel technical ideas inherent in GIS, because these have dominated the field over the past three decades as researchers have worked to make GIS technically feasible and cost effective. In the future, the ideas that emerge from the technology are likely to have a stronger social flavor, as we begin to see the impact of GIS on society at large. Much has already been written on the power of information to change society, and a small part of this literature is specifically concerned with geographic data (Obermeyer and Pinto 1994; King and Kraemer 1993). In the future, we can expect a strong tie to national efforts to coordinate the creation and use of geographic information, particularly the National Spatial Data Infrastructure (National Research Council 1993), to which the Clinton administration appears strongly committed.

But these future prospects aside, the GIS tool and its relatives have already provoked their share of new thoughts and new ways of thinking about the geographical world. Some of these will be enlightening, while others may be as constraining as pencil and paper. After all, to the person with nothing but a hammer, everything sooner or later begins to look like a nail. Similarly, an overly simple GIS can encourage a restricted view of a world in which places are reduced to points, lines, and areas. Only by working to improve the tool can we ensure that its effects continue to be stimulating.

Notes

The National Center for Geographic Information and Analysis is supported by the National Science Foundation, under Cooperative Agreement SBR 8810917.

1. Introductions to geographic information systems are provided by Star and Estes (1990), Burrough (1986), Laurini and Thompson (1992), and many others. Maguire, Goodchild, and Rhind (1991) provide a comprehensive review of the technology.

2. These numbers refer to the ratio between distances on the map and distances on the earth's surface. For example, one inch on a 1:250,000 map equals 250,000 inches on the ground.

3. Suppose, for example, you were studying the relationship between precipitation and elevation in Vermont, and your reporting units are towns. Traditional statistical methods, such as calculating the correlation between precipitation and elevation, assume that observations are independent, that is, the precipitation and elevation in one town are unrelated to those for other towns. Yet we know that a town's elevation and precipitation levels are, in fact, more closely related to levels in nearby neighboring towns than to those in far away towns. The towns should therefore not be treated as independent observations.

References

Abler, R. F. 1988. Awards, Rewards, and Excellence: Keeping Geography Alive and Well. *The Professional Geographer* 40(2):135–140.

Anselin, L. 1989. *What Is Special about Spatial Data? Alternative Perspectives on Spatial Data Analysis.* Technical Paper 89–4. Santa Barbara, Calif.: National Center for Geographic Information and Analysis.

Burrough, P. A. 1986. *Principles of Geographical Information Systems for Land Resources Assessment.* Oxford: Clarendon.

Buttenfield, B. P., and R. B. McMaster, eds. 1991. *Map Generalization: Making Rules for Knowledge Representation.* London: Longman.

Chernoff, H. 1973. The Use of Faces to Represent Points in k-dimensional Space Graphically. *Journal of the American Statistical Association* 70:548–554.

Chui, C. K. 1992. *An Introduction to Wavelets.* Boston: Academic Press.

Daratech. 1994. New GIS Market Study. *Geodetical Info Magazine* (May):36–37.

Department of the Environment. 1987. *Handling Geographic Information: Report of the Committee of Enquiry Chaired by Lord Chorley.* London: Her Majesty's Stationery Office.

Estes, J. E., and D. W. Mooneyhan. 1994. Of Maps and Myths. *Photogrammetric Engineering and Remote Sensing* 60(5):517–524.

Fotheringham, A. S., and P. A. Rogerson, eds. 1994. *Spatial Analysis and GIS.* London: Taylor and Francis.

Goodchild, M. F. 1988. Stepping Over the Line: Technological Constraints and the New Cartography. *The American Cartographer* 15(3):311–319.

Goodchild, M. F., and S. R. Yang. 1992. A Hierarchical Spatial Data Structure for Global Geographic Information Systems. *Computer Vision, Graphics, and Image Processing: Graphical Models and Image Processing* 54(1):31–44.

Goodchild, M. F., B. Klinkenberg, and D. G. Janelle. 1993. A Factorial Model of Aggregate Spatio-temporal Behavior: Application to the Diurnal Cycle. *Geographical Analysis* 25(4): 277–294.

Haining, R. P. 1990. *Spatial Data Analysis in the Social and Environmental Sciences.* New York: Cambridge University Press.

Harley, J. B., and D. Woodward, eds. 1987. *The History of Cartography.* Chicago: University of Chicago Press.

King, J. L., and K. L. Kraemer. 1993. Models, Facts, and the Policy Process: The Political Ecology of Estimated Truth. In *Environmental Modeling with GIS,* eds. M. F. Goodchild, B. O. Parks, and L. T. Steyaert, 353–360. New York: Oxford University Press.

Langran, G. 1992. *Time in Geographic Information Systems.* London: Taylor and Francis.

Laurini, R., and D. Thompson. 1992. *Fundamentals of Spatial Information Systems.* San Diego: Academic Press.

Leick, A. 1990. *GPS Satellite Surveying.* New York: Wiley.

MacDougall, E. B. 1992. Exploratory Analysis, Dynamic Statistical Visualization, and Geographic Information Systems. *Cartography and Geographic Information Systems* 19(4):237–246.

Maguire, D. J. 1991. An Overview and Definition of GIS. In *Geographical Information Systems: Principles and Applications,* eds. D. J. Maguire, M. F. Goodchild, D. W. Rhind, 9–20. London: Longman.

Maguire, D. J., M. F. Goodchild, and D. W. Rhind, eds. 1991. *Geographical Information Systems: Principles and Applications*. London: Longman.

Mandelbrot, B. B. 1982. *The Fractal Geometry of Nature*. San Francisco: Freeman.

Mark, D. M. 1990. Neighbor-based Properties of Some Orderings of Two-dimensional Space. *Geographical Analysis* 22(2):145–157.

Mark, D. M., and A. U. Frank, eds. 1991. *Cognitive and Linguistic Aspects of Geographic Space*. Dordrecht: Kluwer.

Morton, G. M. 1966. *A Computer Oriented Geodetic Data Base and New Technique in File Sequencing*. Ottawa: IBM Canada.

Muehrcke, P. C., and J. O. Muehrcke. 1992. *Map Use: Reading, Analysis, and Interpretation*. Madison: JP Publications.

National Research Council. 1993. *Toward a Coordinated Spatial Data Infrastructure for the Nation*. Washington, D.C.: National Academies Press.

Obermeyer, N. J., and J. K. Pinto. 1994. *Managing Geographic Information Systems*. New York: Guilford.

Openshaw, S. 1983. *The Modifiable Areal Unit Problem*. Norwich: Geobooks.

Openshaw, S., M. Charlton, C. Wymer, and C. Craft. 1987. A Mark I Geographical Analysis Machine for the Automated Analysis of Past Data Sets. *International Journal of Geographical Information Systems* 1:335–358.

Pickles, J. 1995. *Ground Truth: The Social Implications of Geographical Information Systems*. New York: Guilford.

Samet, H. 1990a. *Applications of Spatial Data Structures: Computer Graphics, Image Processing, and GIS*. Reading, Mass.: Addison-Wesley.

———. 1990b. *The Design and Analysis of Spatial Data Structures*. Reading, Mass.: Addison-Wesley.

Smith, N. 1992. History and Philosophy of Geography: Real Wars, Theory Wars. *Progress in Human Geography* 16(2):257–271.

Star, J. L., ed. 1991. *The Integration of Remote Sensing and Geographic Information Systems*. Bethesda: American Society for Photogrammetry and Remote Sensing.

Star, J. L., and J. E. Estes. 1990. *Geographic Information Systems: An Introduction*. Englewood Cliffs, N.J.: Prentice-Hall.

Tomlinson, R. F., H. W. Calkins, and D. F. Marble. 1976. *Computer Handling of Geographical Data*. Paris: UNESCO.

Tukey, J. W. 1977. *Exploratory Data Analysis*. Reading, Mass.: Addison-Wesley.

Van Oosterom, P.J.M. 1993. *Reactive Data Structures for Geographic Information Systems*. New York: Oxford University Press.

Wood, D., with J. Fels. 1992. *The Power of Maps*. New York: Guilford Press.

PART 2

THE WORLD AS
HUMAN HOME

4

Human Adjustment

Robert W. Kates

The concept of human adjustment, applied fifty years ago to flood-plains and now to global change, has served as a practical guide to action, as research paradigm, and as aspiration for humane coexistence with the natural world. The concept is rooted in antiquity as "art in partnership with nature" (Glacken 1967:147), but it was first given its modern expression by Gilbert F. White.

White came to Washington, D.C., in 1934, having completed all but the dissertation for a doctorate at the University of Chicago. His task was to "assist a committee of the Public Works Administration in preparing a comprehensive plan for the Mississippi valley, nothing less. The land and water were to yield more bountifully, drought damage was to be reduced, electricity would flourish, and flood damage would be curbed—the last a goal about which I soon became skeptical" (White 1994:3).

It was this skepticism that flood damages could be curbed by the prevailing technologies of dam, levee, and channel construction that turned White to a broader consideration of how society might adjust to recurrent flood hazard. He did this in what might have been the most influential dissertation ever written in U.S. geography. Begun in 1938 and completed in 1942, he inserted the last footnotes just days before boarding a boat that would take him to Lisbon and to occupied France and four years of refugee service with the American Friends Service Committee. In his dissertation, *Human Adjustment to Floods: A Geographical Approach to the Flood Problem in the United States,* he defined adjustment as an ordering of occupance, or "the human process of occupying or living in an area and the transformations of the initial landscape which result" (White 1945:46). Never comfortable with abstractions, White went on to specify at least eight forms of human adjustment to floods: elevating land, abating floods by land treatment, protecting against floods by levees and dams, providing emergency warning and evacuation, making structural changes in buildings and transportation, changing land use to reduce vulnerability, distributing relief, and taking out insurance.

He then concluded that: "If the resources of the floodplains of the United States are to be used in the public good so as to yield maximum returns to the nation with minimum possible social costs," (White 1945:205) four essential principles should be recognized. First, public policy should take into account all possible adjustments. Second, it should recognize that adjustments are not neutral but rather can favor one form of floodplain use over others. Therefore, third, society should consider carefully the various uses of the floodplains made possible by such adjustments, recognizing the differential need for floodplain use and location. And fourth, society should weigh the full range of social costs and benefits it incurs in employing these adjustments, not merely the costs and benefits that are easy to measure.

None of the foregoing principles had really been applied in 1942, except in the suggestive and exploratory fashion of his dissertation. Thus after returning from overseas service and a stint at a college presidency, White, as professor of geography at the University of Chicago, launched a fifteen-year effort to apply these principles. He sought to identify the various uses of the floodplains across the United States and the adjustments that make possible their occupance (Burton 1962; White et al. 1958), to identify the range of possible adjustments (White 1964) and people's knowledge of them (Kates 1962), to explore in depth the potential for promising but underused adjustments (Murphy 1958; Sheaffer 1960), and to compare the full range of social costs and benefits in particular places (White 1964). From this collective effort, a new public policy and a research paradigm were born.

Practical Applications

The new public policy was enunciated in a congressional document (U.S. Congress 1966), the product of a Bureau of the Budget Task Force that White chaired. Within a decade it had become part of a unified national program for floodplain management (U. S. Water Resources Council 1979). In this program the excessive utilization of flood protection and relief as the major available adjustments was replaced by a broad set of alternatives. Floodplain insurance was made widely available and conditioned on the enactment of floodplain zoning. The insurance program required improved hazard information, and maps of floodplain location were produced for thousands of communities. Floodproofing practices became almost routine in the design of new buildings located in floodplains. Today, a broad set of human adjustments to floods is both conventional wisdom and public policy, although the implementation of these adjustments lags behind its potential and the pressures for continued expansion onto floodplains and coastal zones (U.S. Federal Interagency Floodplain Management Task Force 1992; U.S. Interagency Floodplain Management Review Committee 1994). From our professional point of view, the application of White's idea of human adjustment is the best example of a public policy success story in which a geographical concept, strongly supported by extensive empirical research, changes the ways things are done (Platt 1986).

In the United States and internationally, a broadened choice of adjustments has been applied to other hazards including avalanches, coastal flooding, coastal erosion, droughts, earthquakes, hurricanes, snow, tropical cyclones, volcanic activity, and wind as well as the hazards of a place. The approach has been extended to technological hazards, beginning with studies of air pollution and later including automobiles, nuclear power, airborne mercury, PCBs, consumer products, contraceptives, and even television. (For details on these applications, see reviews in Alexander 1991; Burton, Kates, and White 1993; Cutter, 1993; Mitchell 1989; O'Riordan 1986; Palm 1990; Whyte 1986).

In resource management, the concept of human adjustment has enlarged alternatives in water resource development (White 1969; Day, et al. 1986). The concept has also led to improved strategies for providing water supply and sanitation in developing countries (White, Bradley, and White 1972); it has enabled the spread of opportunities to conserve energy (Boulding 1986) and led to many experiments in reducing or reusing solid waste and industrial materials (Ausubel and Sladovich 1989); and it has aided efforts to conserve food and food crops (Tait and Napompeth 1987).

Recurrent themes mark all of these examples. Prevailing practice always tends to be narrow and to favor technological solutions over behavioral ones. Such practices seek to control natural events, to increase resource supplies, or to clean up waste and pollutants rather than to reduce hazard vulnerability, resource need, or streams of waste and pollutants. The successes involve broadening the sets of available adjustments and melding behavioral and technological approaches, thus raising the threshold of hazard vulnerability, reducing needs for energy or water, and diminishing the streams of wastes and pollutants. As California coped with drought, Rhode Island recycled statewide, and public utilities everywhere encouraged energy conservation, knowledge of the range of alternatives became widespread and the choice of alternatives was more and more a matter of public debate and advocacy. And underlying each and every one of these successes has been a significant body of research.

Research Paradigm

Research paradigms are not consciously created; they only become, and this one was appropriately announced in a mimeographed working paper (Burton, Kates, and White 1968) that suggested to geographers and others interested in natural hazards research five areas on which to collect comparable data: the extent of occupance in areas subject to natural hazard, the range of adjustments found in such areas, their residents' or users' perceptions of the hazard, their process of choice, and the influence, if any, of public policies. The working paper served as an initial document for a major collaborative research effort under the aegis of the International Geographical Union. Geographers in fifteen countries and forty sites applied the paradigm to the occupance of areas affected by nine different hazards. About five thousand people worldwide who used the resources or resided in such hazard zones were interviewed to learn how they adjust to the hazards that threaten them. White's eight adjustments were found to have their universal counterparts: everywhere people bear or share hazard losses, everywhere people try to modify hazard events or prevent their effects, and some will even change their resource use or move to new locations (White 1974; Burton, Kates, and White 1978).

But the richness of human experience found in these international comparisons revealed another form of adjustment embedded in the larger fabric of social life and human activity. Besides the purposeful adjustments, all societies seemed to employ adjustments that have the effect of reducing or mitigating the hazard but were not consciously invoked for that purpose (Figure 4.1). Rather, these adjustments were incidental to some other activity with another intended function. For example, the choice of a brick house, the prevailing style in the Midwest, rather than a wooden one also reduces the potential for tornado damage. Or the universal diffusion of television and radio can also improve hazard warning. Still other adjustments were embedded

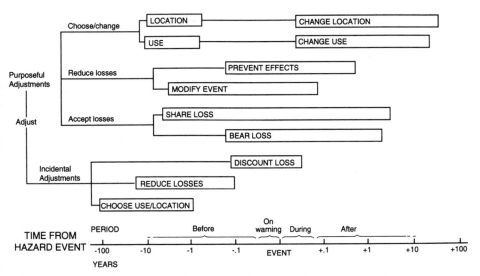

Figure 4.1. A Choice Tree of Adjustment. Adjustment begins with an initial choice of resource use, livelihood system, and location. For that set of choices, various incidental and purposeful adjustments are available at somewhat different time scales for initiation. The most radical choice is to change the original use or location. Source: Burton, Kates, and White (1978:46).

deeply in the culture (in a few cases in biology), such a part of everyday life and practice that they were inseparable from the context, settings, or givens of a particular society, place, hazard, or resource.

Over time, adjustment has evolved into a generic concept (Burton, Kates, and White 1978; Mileti 1980). "Adjustment" becomes synonymous with the short-term, the conscious, and the purposeful. The term adaptation has come to refer to the long-run adjustments successfully integrated into culture, or in very rare cases, even into biological evolution (see Table 4.1). But maladjustment is also recognized (Whyte 1986); practices intended to reduce vulnerability may increase it in the long run, and the evolution of cultures often increases some types of vulnerability.

Interdisciplinary Developments

The years that saw the emergence of this research paradigm in which human adjustment is a central feature also saw the development of two major interdisciplinary themes that still elicit widespread interest and study. The first was the emergence of systems as a framework for analyzing how the world works (Bertalanffy 1956). Incorporating concepts from electrical engineering and biology, general systems theory

Table 4.1. Human Adjustments to the Risk of Environmental Extremes

Purposeful Adjustments

Choose/Change	*Reduce Loss*	*Redistribution of Loss*
Change locations • Abandonment	*Modify event* • Weather modification	*Share losses* • Insurance • Disaster relief • Charity
Change use • Land use planning	*Affect cause* • Hazard specific	*Bear loss* • Create reserve funds for anticipated loss
	Prevent losses • Warning systems • Building codes • Engineering works • Evacuation	

Incidental Adjustments

Choose Use/Location
• Land use regulation,
  nonhazard specific
	Reduce Loss
	• Fire codes
	• Transportation
	  improvement
	• Fire fighting
	  improvement
		Redistribution of Loss
		• Savings

Unwitting Adaptations

Biological • Undetermined	*Cultural* • Deurbanization (sprawl) • Shifts in aesthetic preferences • Shifts in family structure • Reduction in wealth

SOURCE: Mileti 1980.

sought to explain the linkages between phenomena and their resulting organization at scales ranging from atoms to the universe. This emergent quasi-discipline created broad excitement, popularized new terminology, and was eventually incorporated into the everyday practice of science so that one can hardly remember when models, inputs and outputs, and feedback did not exist. But to those of us who became addicted to drawing boxes to represent phenomena and arrows to indicate the flows between them, nature and society deserved equal boxes, and human adjustment was an obvious form of negative feedback, reducing the effects of the extremes of nature much as a thermostat keeps a house at an even temperature (Kates 1971). Hazard researchers contributed as well, providing relatively well-worked out examples of complex systems that included nature and society, with multiple adjustments as sources of feedback.

Early on, White recognized that some effective short-run human adjustments might actually increase long-run vulnerability. His favored example was the levee effect—the observation that levees serve well to prevent flood damages, but if one waits long enough, all levees are overtopped, so that the wall that kept the water out becomes the wall that keeps the water in, thereby deepening the ensuing catastrophe (White et al. 1958). From our observations of the roles of human adjustment, we derived three generic hypotheses (Burton, Kates, and White 1978; Bowden et al. 1981). The *lessening* hypothesis stated that over time, the thrust of development in societies and nations is toward reducing the social costs of hazard to society, especially deaths, by broadening the range of effective adjustments. Thus, although the number of recorded disasters worldwide in which more than one hundred died has doubled in the second half of this century compared to the first, the total death toll has decreased considerably (Burton, Kates, and White 1993:11). The *transition* hypothesis states, however, that in periods of rapid social and economic change, older forms of adjustment may deteriorate before newer forms become available; societies become peculiarly vulnerable to hazard at such times. Thus in many developing countries, traditional systems of sharing in times of disaster have eroded because of migration, urbanization, and commercialization while the substitute systems of organized relief and welfare are still poorly developed. The *catastrophic* hypothesis states that successful lessening of hazard may serve to increase a population's vulnerability to a catastrophic perturbation that exceeds the level of adjustment. For example, deltaic islands in Bangladesh that were partially protected by artificial embankments encouraged land-poor people to farm these areas, many of whom died when the embankments were subsequently overtopped by a major storm. This combination of multiple case studies and generic hypotheses made for exciting conversations among geographers, ecologists, engineers, mathematicians, and physicists about the evolution of systems theory into nonlinear dynamics, catastrophe theory, surprise, and the current buzzword of chaos (Holling 1973; Johnson and Gould 1984; Svedin and Aniansson 1987).

The second fruitful interdiscipline emerged under the rubric of decision science, or how people or institutions do or should make decisions. Again we were able to draw upon and use new concepts of risk and probability (Savage 1954), decision making under uncertainty (Edwards 1954), and bounded rationality (Simon 1957), choosing particularly those concepts that seemed to broaden the narrow focus of neoclassical economics with its assumptions of almost perfect knowledge and utility-maximizing behavior. And again we were able to offer examples of people making choices of adjustments under real conditions of uncertainty (Kates 1962) rather than the contrived choice making of students in introductory psychology classes. These initial collaborations with psychologists (Sims and Baumann 1972; Slovic, Kunreuther, and White 1974) would help to spin off other vital fields— behavioral geography and environmental psychology—and to enable geographers to take an important role in another emerging interdiscipline, that of risk assessment and analysis (White 1972).

The emphasis on comparable empirical studies of adjustment, the employment of systems concepts, and the focus on choice provoked major critiques of the paradigm (for a review see Chapter 9 in Burton, Kates, and White 1993). In retrospect some of the critique of the paradigm was obviously parochial, reflecting from within geography the long-standing tension between the focus either on the particular or on the universal, and from outside, the varied assertions of disciplinary turf. More important, and for understandable reasons, the critiques invariably lagged behind the moving front of inquiry and therefore often seemed dated. Nonetheless, two lines of criticism had considerable validity.

Critiques of the Paradigm

The first criticism asserted that the use of floods and other extreme geophysical events as a starting point for analysis was too narrow, too deterministic, or both. The vulnerabilities of people are rooted in the precariousness of everyday existence, not necessarily in the rare and extreme event. In real life, the everyday and the extraordinary are intermixed. It is collectivities of threats—natural, social, and technological—that threaten human existence, and the salience of any one changes over time. Thus although the interactive systems notions of nature and society with feedback might be an improvement over the simplistic determinism of a linear model of natural cause and human loss, these notions were seen as still excessively simplistic, ignoring the complexity of social reality.

The second critique arose primarily in response to the emphasis on choice and decision, with their implications that humans are masters of their fate. The emphasis on choice of adjustment seemed to ignore the reality of constraint and structure that is part of social existence. It also focused on the microdecision, often ignoring the larger social structures and their dynamics that bind and constrain the decisions people make. This emphasis appeared to be rooted in an ideology of individualism that ignored the reality of what was at best a constraint and at worst the oppression that characterized much social experience. Though in theory it might be possible to choose from a broad range of adjustments, the choice was actually denied to many by virtue of their life circumstance, social class, or income.

Considerable progress has been made in addressing these two concerns. Human adjustment studies have now spanned an enormous range of natural, technological, and social hazards and resources and are beginning to tackle even larger phenomena such as hunger or environmental degradation (for a review see Chapter 9 in Burton, Kates, and White 1993). Models of adjustment have moved in two directions. On the one hand they have become more precise as models of causality where chains of events, consequences, and adjustments are clearly specified and linked. On the other, they have expanded as various attempts are made to link the everyday and the extraordinary and to link microdecisions with macrostructures. In all these efforts, considerations of differential vulnerability and of restricted access to adjustments are standard issues in analysis—issues of great importance as we consider adjusting to the great global changes currently underway.

Adjusting to Human-induced Change

The ultimate challenge of adjustment is to global human-induced environmental change. In 1987 an international colloquium took stock of the magnitude of human induced change (the contributions, by Turner et al., were published in 1990 as *The Earth As Transformed by Human Action*). A sampling of their findings show that since the dawn of agriculture ten thousand years ago, an area the size of the continental United States has been deforested by human effort. Today, half of the ecosystems of the ice-free lands of the earth have been modified, managed, or utilized by people. The flows of materials and energy that are removed from their natural settings or synthesized now rival the flows of such materials within nature itself. Water, in an amount greater than the contents of Lake Huron, is withdrawn each year for human use. All told, the Earth Transformed project was able to reconstruct human-induced change in thirteen worldwide measures of chemical flow, land cover, and biotic diversity. If we consider the total amount of human-induced change represented by these thirteen measures over the last ten thousand years, then most of it has been extraordinarily recent, with seven of the measures having doubled since 1950.

The rapidity of these changes lends credence to the fears for the earth's fate—fears enhanced by the changes yet to come (Kates 1994). For it is over the next sixty to eighty years that forecasters foresee a doubling of the earth's population. To meet the needs of this population, some project a four-fold increase in agriculture, a six-fold rise in energy use, and an eight-fold increase in the value of the global economy (Anderberg 1989). The environmental changes caused by such enormous intensification of production and consumption could well be too much for human health, habitat, and well-being, and for the life support systems of nature and the earth.

These threats to the earth are many and varied. If one adopts criteria that consider which are the most likely to occur, to cause the most harm, or to affect the most people, three groups of large-scale and long-term threats have emerged. The first are the *global atmospheric* concerns of nuclear proliferation, stratospheric ozone depletion, and climatic warming from greenhouse gasses. The second are massive *assaults on the biota*, specifically deforestation in the tropical and mountain lands, desertification in the drylands, and species extinction, particularly in the tropics. The last are the large-scale introductions of *pollutants*—things such as acid rain in the atmosphere, heavy metals accumulating in the soils, and chemicals in the groundwater.

Preservationists versus Adaptationists

An active debate now focuses on how to respond to the profound changes underway. The debate, particularly that over global atmospheric concerns, has been characterized by its opposite poles of "preventionist" versus "adaptationist" positions (Mathews 1987). The discussion is confused a bit by the terms used, with the current international terminology distinguishing between "mitigation," implying a prevention or reduction in climate changes, and "adaptation," describing both purposeful and incidental adjustment to such changes as well as unwitting adaptation. But whatever the terminology, preventionists argue that the impacts of these changes on humankind, particularly the accumulation of greenhouse gasses, are potentially so catastrophic that mitigation actions are needed now to drastically slow or reduce the rate of change (Mathews 1987). Adaptationists counter by noting the relatively slow process of climatic change that is forecast and by arguing that human society can surely adapt to such changes. They cite the ubiquity of human settlement in all climates, note that the climate changes that people encounter in the course of migration exceed any of those forecast for the earth's future, and point to the continuing process of adapting desired plants and agriculture well beyond their original settings (Ausubel 1991).

Many who study global change, particularly from industrialized countries, are optimistic about adaptive capacity. For example, they are confident that agricultural enterprise and technology can successfully adjust to massive environmental change and will be able to provide the large amounts of food that will be needed in the future. Their confidence is based on historic trends in yield increases, on the spread of cropping systems far beyond their traditional agroecological boundaries, and on the inherent flexibility of systems of international trade (NAS-NAE-IOM 1992).

Although adaptation clearly plays a critical role, the possibilities for adaptation are surely different in rich and poor countries (Bohle, Downing, and Watts 1994). Because adaptation, even by the invisible hand of the market, is not cost-free and does not yield the same benefits everywhere, we need to understand much more about the social costs of adaptation and differential access to it. Evaluating the impacts of environmental change requires assessing the full social costs of adaptation, including the secondary effects of the adaptations themselves and the losses suffered by the groups and locations bypassed or marginalized by the ensuing changes.

Unfortunately, there has been little direct study of such social costs, the secondary effects, and the losses from failure to adjust. One can speculate on why this has been the case, but I think there is a reluctance to undertake such study. The preventionists are biased against studies of adaptation, fearing that such work may weaken the social will to undertake greenhouse gas reduction and will play into the hands of those who argue that any action is premature. Many adaptationists see no need to study the social costs of adaptation; they either ignore the issue, thinking that the costs are trivial, or simply trust the invisible hand of market forces. But designing good studies is also difficult. Thus, much of the reasoning as to the success (or in rare cases, failure) of adjustment is by analogy; it comes from observing how people have adjusted to instances of socioeconomic or technological change or to short-term environmental change (Glantz 1988).

In this debate, many geographers find themselves awkwardly perched on the fence line. As professionals, we are surely unmatched in our concern for the fate of the earth, and few are more knowledgeable of the extraordinary adaptability of human livelihood systems. In the echoes of time, we hear the preservationists—good environmentalists all—sound a bit like the Army Corps of Engineers of fifty years ago, but instead of trying to stem flood waters, the focus this time is on the stream of carbon flowing into the atmosphere. We also hear in the adaptationists echoes of the neoclassical lawyers and economists who argued that floodplain zoning was an unnecessary restriction on property rights and that the market could shape the needed floodplain adjustments.

To this ongoing debate, we geographers can offer fifty years of theoretical insight, empirical research, and practical experience. From our middle, somewhat uncomfortable ground, I suspect that we will be carefully monitoring the rates of change and will call for their slowing to provide time for adjustment. Yet we will also assert our faith in the human ability to adapt while helping to identify a broad range of alternative adjustments. And perhaps most important, we will serve as advocates for those who are most vulnerable to the environmental changes and who have the least access to the needed adjustments.

Research on Differential Adjustment

Two sets of recent studies, partly organized by geographers, provide research needed to support such advocacy. They address the differential ability to adjust—one from a global perspective using nations or regions as the units of analysis, the other local—based on case studies of poverty and environment.

The first study explores the impact of future climate change on food security. It brings together the work of agriculturists, economists, geographers, and others concerned with the supply and demand for food with the work of those involved in assessing climate change and its implications for agricultural systems (Rosenzweig and Parry 1994). This state-of-the-art study found that on average, climate change would have only a small impact overall on global agricultural productivity, assuming a modest level of adjustment and the enrichment benefits for plant growth of increased carbon dioxide in the atmosphere.

This study utilized a set of thirty-five country and country-group agricultural policy models designed to simulate the national or regional agricultural sectors that are linked together through trade, world market prices, and financial flows. The models were used to generate a "reference" scenario for the year 2060, which projects that economic activity will increase by a factor of 4.4 over 1980 levels and that cereal production will increase by a factor of 2.25 over this period, barely keeping up with population growth. Utilizing the relationship between per capita food availability and nutritional requirements and assuming no change in income distribution, the number of undernourished would rise from 501 million in 1980 to 641 million in 2060. This increase, however, would represent a significant decline in the *proportion* of undernourished, from 23 percent to 9 percent of the world's population.

Rosenzweig and Parry (1994) created a second set of scenarios to test the sensitivity of the reference scenario to a changing climate using three different general circulation model (GCM) simulations under conditions of doubled atmospheric concentrations of carbon dioxide (CO_2). They then used the simulated climate data as inputs into crop models for wheat, maize, rice, and soybeans applied to more than one hundred sites in eighteen countries. The crop models take into account the effects of higher CO_2 concentrations on plant growth and water use and also allow for effects of adjustments in the form of farm management practices. Two levels of adjustment (adaptation) are employed: those to be undertaken at the farm level without major changes in the agricultural system and those requiring a transformation of the agricultural system itself. The resulting yield changes for three sets of GCM runs and four different cases (with and without the CO_2 fertilization effect and with two levels of adaptation) were aggregated regionally and nationally and used as inputs into the system of linked agricultural models.

Figure 4.2 graphs the resulting changes in cereal production in 2060 for each of three different climate models with CO_2 enrichment (on the x-axis) and with two levels of adaptation. On balance, as shown by the black bars for the globe as a whole, production with adaptation is reduced by less than 3 percent and even increases in one of the climate scenarios with a high level of adaptation. This global average is sharply differentiated, however, between developed and developing countries. In all cases, climate change leads to net declines in cereal production in the developing world. Increases in production in the developed world in most instances do not fully counterbalance the developing world declines, except at high levels of adaptation. In the simulation, world market prices increase at the same time that developing countries are forced to import more cereals. As a result, the estimated number of undernourished people is projected to increase in eleven of the twelve climate change scenarios examined, increasing to almost two billion people in the most extreme case.

One of the early results of hazard research is that people's ability to adjust and their access to adjustments reflect existing divisions between rich and poor, powerful and powerless, ethnic or gender-favored and ethnic or gender-denied. Wisner (1977), for example, identified such differences in drought adjustment in Kenya. Subsequent case studies on the linkages between poverty and environmental degrada-

Change in Cereal Production in 2060

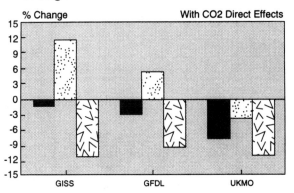

 Global

 Developed

Developing

Adaptation 1

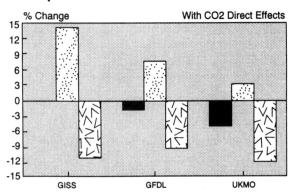

Adaptation 2

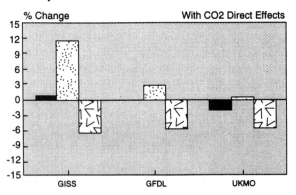

tion have enhanced our view of the complex ways adjustment is denied to poor people. Viola Haarmann and I collected a set of thirty case studies in an effort to document the relationship between poverty and environment (Kates and Haarmann 1992). Despite the widespread view that poverty and environmental degradation are strongly linked, relatively few studies carefully document the actual linkages. For the rural inhabitants of these case study locales, maintaining access to the natural resource base and inputs for agriculture, herding, or fishing is becoming increasingly difficult in the face of growing population, increased competition for land, and "development." Thus the available case studies tell common tales of poor people's displacement from their lands, the division of their resources, and the degradation of their environments.

The poor are *displaced* by activities that, in the name of development or commercialization, deprive them of their traditional entitlement to the common property resources so essential to their survival. The entitlements of the poor are *divided* and reduced by their need to share their resources with their children or to sell off bits and pieces of their resources to cope with extreme losses (crop failure, illness, death), social requirements (marriages, celebrations), or simple subsistence. The resources of the poor are *degraded* by excessive or inappropriate use, by failure to restore or to adequately maintain protective works, and by the loss of productive capacity from natural hazards.

These processes are driven by four major forces that culminate in a spiral of impoverishment and degradation (Figure 4.3). Two are forces external to the case study locales: development/commercialization and natural hazards. Two are internal to the communities studied: population growth and existing poverty. The case studies tell of three major sequences of spirals of impoverishment and degradation, in each of which two of the driving forces dominate. In the first sequence, driven by development activities or commercialization and by population growth, poor people are *displaced* from their resources by richer claim-

Figure 4.2. Change in Cereal Production in 2060. Three climate models (GISS, GFDL, and UKMO) were used to simulate climate changes with a doubling of CO_2 in the atmosphere. These are linked to crop models for maize, wheat, and rice that include the favorable direct effects of CO_2 on yield. The impacts of climate change in 2060 on cereal production are shown as the percentage change from the reference scenario (without climate change) globally, and in developed and developing countries. In Figure 4.2's top chart, no adaptation (adjustments) are assumed, in the center chart, Adaptation 1, minor farm-level adjustments are assumed; in the bottom chart, Adaptation 2, major changes in agricultural systems are assumed. Source: Rosenzweig and Parry (1994:137).

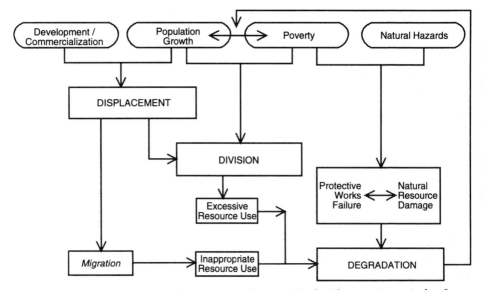

Figure 4.3. Impoverishment-Degradation Spirals. Three major spirals of household impoverishment and environmental degradation are driven by combinations of driving forces of development and commercialization, population growth, poverty, and natural hazards. Source: Kates and Haarmann (1992:9).

ants or by the competition for existing land or employment. The displaced then face division of the remaining resources or are forced to migrate to other, usually more marginal, areas. In the second sequence, driven by population growth and existing poverty, meager resources are further *divided* to meet the needs of new generations or the crises arising from their poverty. Resources are then *degraded* by excessive use of divided lands or by inappropriate use of environments unable to sustain the requisite resource use. In the third sequence, even in the absence of further displacement and division, poverty creates poor households unable to maintain needed protective works or to restore resources, while natural hazards like disease, drought, flood, soil erosion, landslides, and pests further degrade these natural resources.

These exogenous driving forces frequently involve human adjustments, which while benefitting some may harm poor people or are not accessible to poor people because of their poverty or powerlessness. Thus many development-commercialization activities—large-scale agriculture, irrigation, hydroelectric development, forestry, and wildlife preservation activities—which are in themselves major adjustments to environmental constraint or opportunity, do benefit some but frequently displace poor people. Poverty also restricts people's ability to maintain existing adjustments because they lack the labor to restore or to maintain protective works (terraces, drainage, etc.), the means to hire specialized skills or make needed inputs, or the access to public programs of resource improvement and renewal. Recurring natural hazards require continuous efforts to adjust, with pauperization and enforced migration resulting from failure to adjust. Taken together, these studies tell us that the social costs of adjustment include the displacement of poor people from their lands, waters, and vegetation; that poor people are hard-pressed to maintain even existing adjustments; and that the failure to adjust incurs a heavy penalty on those least able to bear additional burdens.

Coexistence with Nature

For fifty years the concept of human adjustment has fostered a powerful research paradigm. It has had, and continues to have, practical application, now especially in the context of the great global changes currently underway. But its most powerful influence may be heuristic, encouraging our coexistence with the natural world.

Such coexistence is impossible, McKibben (1989) asserts, for *The End of Nature* has already occurred, its demise created by just possessing the knowledge of how humans have changed nature. Such coexistence is unnecessary, Lovelock (1990) implies in his Gaia hypothesis, because we humans are but a part of an immense self-regulating biogeochemical system, already eons old before our appearance on earth. Indeed we are both *part* of and *apart* from the natural world. Because we are human and have the gift of consciousness, we are forever locked into the dualities in being a part of our world—living human organisms dependent for our existence on the natural world—and apart from it by virtue of our consciousness and reflection.

Adjustment—purposeful, incidental, or unwitting—seeks to bridge these troubled waters. In our geographic tradition, it draws on a "possibilism," in which characteristics of the natural world may encourage or constrain particular human uses of the earth, but never determine them. Rather, through human adjustment and adaptation to an enormous range of natural settings, virtually all of the earth has become a place of residence and a source of sustenance for people. Human adjustment is clearly an anthropocentric concept—human welfare is the measure of things. But it is a position that advocates coexistence, even coevolution, with nature, rather than human dominance of nature. Adjustments that do not take into account the distinctive characteristics of environment are costly, and in the long run unsustainable. Thus for we humans—children of Gaia, the mother of all—human adjustment is what we do while growing up.

References

Alexander, D. 1991. Natural Disasters: A Framework for Research and Teaching. *Disasters* 15(3):209–226.

Anderberg, S. 1989. A Conventional Wisdom Scenario for Global Population, Energy, and Agriculture 1975–2075, and Surprise-rich Scenarios for Global Population, Energy and Agriculture 1975–2075. In *Scenarios of Socioeconomic Development for Studies of Global Environmental Change: A Critical Review*, ed. F. L. Toth et al., 201–279. RR-89-4. Laxenburg, Austria: IIIASA.

Ausubel, J. H. 1991. Does Climate Still Matter? *Nature* 350:649–652.

Ausubel, J. H., and H. E. Sladovich. 1989. *Technology and Environment*. Washington, D.C.: National Academy Press.

Bertalanffy, L. Von. 1956. General System Theory. *Yearbook of the Society for General Systems Research* 1:1–10.

Bohle, H. G., T. E. Downing, and M. J. Watts. 1994. Climate Change and Social Vulnerability: Towards a Sociology and Geography of Food Insecurity. *Global Environmental Change* 4(1):37–48.

Boulding, K. E. 1986. Energy Policy in Black and White: Belated Reflections on a Time to Choose. In *Geography, Resources, and Environment*, vol. 2, *Themes from the Work of Gilbert F. White*, ed. R. W. Kates and I. Burton, 310–325. Chicago: University of Chicago Press.

Bowden, M. J., R. W. Kates, P. A. Kay, W. E. Riebsame, R. A. Warrick, D. L. Johnson, H. A. Gould, and D. Wiener. 1981. The Effect of Climate Fluctuations on Human Populations: Two Hypotheses." In *Climate and History: Studies in Past Climates and Their Impact on Man*, ed. T.M.L. Wigley and G. Farmer, 479–513. Cambridge: Cambridge University Press.

Burton, I. 1962. *Types of Agricultural Occupance of Flood Plains in the United States*. Research Paper 75. Chicago: University of Chicago, Department of Geography.

Burton, I., R. W. Kates, and G. F. White, 1968. *The Human Ecology of Extreme Geophysical Events*. Natural Hazard Working Paper 1. Toronto: University of Toronto Press.

———. 1978. *The Environment as Hazard*. New York: Oxford University Press.

———. 1993. *The Environment as Hazard*, 2nd edition. New York: Guilford Press.

Cutter, S. L. 1993. *Living with Risk: A Geography of Technological Hazards*. London: Edward Arnold.

Day, J. C., E. Fano, T. R. Lee, F. Quinn, and W.R.D. Sewell. 1986. River Basin Development. In *Geography, Resources, and Environment*, vol. 2, *Themes from the Work of Gilbert F. White*, ed. R. W. Kates and I. Burton, 116–152. Chicago: University of Chicago Press.

Edwards, W. 1954. The Theory of Decision Making. *Psychological Bulletin* 51:380–417.

Glacken, C. 1967. *Traces on the Rhodian Shore: Nature and Culture in Western Thought from Ancient Times to the End of the Eighteenth Century.* Berkeley: University of California Press.

Glantz, M. H. 1988. *Societal Responses to Regional Climatic Change: Forecasting by Analogy.* Boulder, Colo.: Westview Press.

Holling, C. S. 1973. Resilience and Stability of Ecological Systems. *Annual Review of Ecological Systems* 4:1–23.

Johnson, D. L., and H. Gould. 1984. The Effects of Climate Fluctuations on Human Populations: A Case Study of Mesopotamian Society. In *Climate and Development*, ed. A. K. Biswas, 117–136. Dublin: Tycooly International.

Kates, R. W. 1962. *Hazard and Choice Perception in Flood Plain Management.* Research Paper 78. Chicago: University of Chicago, Department of Geography.

———. 1971. Natural Hazard in Human Ecological Perspective: Hypotheses and Models. *Economic Geography* 47(3):438–451.

———. 1994. Sustaining Life on Earth. *Scientific American* 271(4):114–122.

Kates, R. W., and V. Haarmann. 1992. Where the Poor Live: Are the Assumptions Correct? *Environment* 34(4):4–11, 25–28.

Lovelock, J. 1990. *The Ages of Gaia: a Biography of Our Living Earth.* New York: Bantam Books.

McKibben, B. 1989. *The End of Nature.* New York: Random House.

Mathews, J. T. 1987. Global Climate Change: Toward a Greenhouse Policy. *Issues in Science and Technology* 3(3):57–68.

Mileti, D. S. 1980. Human Adjustment to the Risk of Environmental Extremes. *Sociology and Social Research* 64(3):327–347.

Mitchell, J. K. 1989. Hazards Research. In *Geography in America*, ed. G. L. Gaile and C. J. Willmott, 410–424. Columbus, Ohio: Merrill.

Murphy, F. C. 1958. *Regulating Flood Plain Development.* Research Paper 56. Chicago: University of Chicago, Department of Geography.

NAS-NAE-IOM Committee on Science, Engineering, and Public Policy, Panel on Policy Implications of Greenhouse Warming. 1992. *Policy Implications of Greenhouse Warming; Mitigation, Adaptation, and the Science Base.* Washington, D.C.: National Academy Press.

O'Riordan, T. 1986. Coping with Environmental Hazards. In *Geography, Resources, and Environment*, vol. 2, *Themes from the Work of Gilbert F. White*, ed. R. W. Kates and I. Burton, 272–309. Chicago: University of Chicago Press.

Palm, R. I. 1990. *Natural Hazards: An Integrative Framework for Research and Planning.* Baltimore: Johns Hopkins University Press.

Platt, R. H. 1986. Flood and Man: A Geographer's Agenda. In *Geography, Resources, and Environment*, vol. 2, *Themes from the Work of Gilbert F. White*, ed. R. W. Kates and I. Burton, 28–68. Chicago: University of Chicago Press.

Rosenzweig, C., and M. L. Parry. 1994. Potential Impact of Climate Change on World Food Supply. *Nature* 367:133–138.

Savage, L. J. 1954. *The Foundations of Statistics.* New York: John Wiley.

Sheaffer, J. R. 1960. *Flood Proofing: An Element in a Flood Damage Reduction Program.* Research Paper 65. Chicago: University of Chicago, Department of Geography.

Sims, J., and D. Baumann. 1972. The Tornado Threat: Coping Styles of the North and South. *Science* 176:1386–1392.

Simon, H. 1957. *Models of Man: Social and Rational.* New York: John Wiley.

Slovic, P., H. Kunreuther, and G. F. White. 1974. Decision Processes, Rationality, and Adjustment to Natural Hazards. In *Natural Hazards: Local, National, Global,* ed. G. F. White, 187–205. New York: Oxford University Press.

Svedin, U., and B. Aniansson, eds. 1987. *Surprising Futures, Notes from an International Workshop on Long-term World Development.* Stockholm: Swedish Council for Planning and Coordination of Research.

Tait, J., and B. Napompeth, eds. 1987. *Management of Pests and Pesticides: Farmer's Perceptions and Practices.* Boulder, Colo.: Westview Press.

Turner, B. L., II, W. C. Clark, R. W. Kates, J. F. Richards, J. T. Mathews, and W. B. Meyer, eds. 1990. *The Earth as Transformed by Human Action: Global and Regional Changes in the Biosphere over the past 300 Years.* Cambridge: Cambridge University Press.

U.S. Congress. 1966. *A Unified National Program for Managing Flood Losses.* 87th Congress, 2nd Session, House Document 465. Washington, D.C.: U.S. Government Printing Office.

U.S. Federal Interagency Floodplain Management Task Force. 1992. *Floodplain Management in the United States: An Assessment Report,* 2 vols. FIA 17–18. Washington, D.C.: U.S. Federal Emergency Management Agency.

U.S. Interagency Floodplain Management Review Committee. 1994. *Sharing the Challenge: Floodplain Management in the 21st Century.* Washington, D.C.: U.S. Government Printing Office.

U.S. Water Resources Council. 1979. *A Unified National Program for Flood Plain Management.* Washington, D.C.: U.S. Water Resources Council.

White, G. F. 1945. *Human Adjustment to Floods: A Geographical Approach to the Flood Problem in the United States.* Research Paper 29. Chicago: University of Chicago, Department of Geography.

———. 1964. *Choice of Adjustments to Floods.* Research Paper 93. Chicago: University of Chicago, Department of Geography.

———. 1969. *Strategies of American Water Management.* Ann Arbor: University of Michigan Press.

———. 1972. Human Response to Natural Hazard. In *Policy Perspectives on Benefit-risk Decision Making,* Committee on Public Engineering, 43–49. Washington, D.C.: National Academy of Engineering.

———. 1994. Reflections on Changing Perceptions of the Earth. *Annual Review of Energy and the Environment* 19:1–13.

White, G. F., ed. 1974. *Natural Hazards: Local, National, Global.* New York: Oxford University Press.

White, G. F., W. C. Calef, J. W. Hudson, H. M. Mayer, J. R. Sheaffer, and D. J. Volk. 1958. *Changes in the Urban Occupance of Flood Plains in the United States.* Research Paper 57. Chicago: University of Chicago, Department of Geography.

White, G. F., D. J. Bradley, and A. U. White. 1972. *Drawers of Water: Domestic Water Use in East Africa.* Chicago: University of Chicago Press.

Whyte, A. V. 1986. From Hazard Perception to Human Ecology. In *Geography, Resources, and Environment*, vol. 2, *Themes from the Work of Gilbert F. White*, ed. R. W. Kates and I. Burton, 240–271. Chicago: University of Chicago Press.

Wisner, B. G., Jr. 1977. *The Human Ecology of Drought in Eastern Kenya.* Ph.d. Dissertation, Clark University, Worcester, Mass.

5

Water Budget Climatology

John R. Mather

Many knowledgeable individuals believe that water will be the next real environmental crisis for the world. Many regions of the world face difficult water problems, including quality and quantity of supply, ownership of water rights, the loss of fresh water sources through misuse, and even the need to create new supplies to meet growing demands. Comprehensive plans based on quantitative knowledge of our water resources are necessary if rational development of these resources is to occur.

The concept of a climatic water budget is fundamental to any evaluation of what might happen as societies, either willfully or inadvertently, undertake both small- or large-scale modifications of their environments. Such a water budget results from a daily, weekly, or monthly comparison of the supply of water from precipitation with the climatic demand for water as given by evapotranspiration (the combined evaporation from water or other moist surfaces and transpiration from vegetative surfaces).

Consider the environmental consequences of replacing a wooded tract with a shopping center or an industrial park. Most states in the United States require that recharge to the local water table after development be unchanged from the before-development recharge. This is a relatively straightforward problem using the climatic water budget. From the bookkeeping steps in the water budget, one can easily compute groundwater recharge from current information on total precipitation and evapotranspiration from the forested area after deducting any overland runoff that might occur. Next, one can determine the groundwater recharge that will occur after land use change by substituting new values for storage in the root zone in the soil, of overland runoff from the new land surface conditions, and even for the increased evapotranspiration resulting from any increased temperature conditions that might develop as a result of the land use change. Comparison with before-development recharge will allow adjustments to be made through the use of detention ponds, storm sewers, and the like to bring pre- and post-development recharge into agreement.

Climatology is concerned with the distribution and use of solar radiation and precipitation in the atmosphere and at the surface of the earth. Water budget climatology is, therefore, one of the core components of the field for it seeks to quantify how precipitation is used for evapotranspiration, overland and subsurface runoff, and stream flow, as well as for recharge and storage of moisture in the upper layers of the soil.

Water budgets may be evaluated over a wide range of time and space scales. For example, the world water budget, better known as the hydrologic cycle, provides annual information on latitudinal, continental, or global values of precipitation, evapotranspiration, and runoff information that is essential to evaluating regional or global water resources. As the space and time scales contract, it is possible to include more detail in a water budget analysis until finally the water budget of a small river basin may provide information on the values of overland runoff, subsurface flow to the groundwater table, soil moisture storage, and evapotranspiration losses from the soil and vegetation on a daily, weekly, or monthly basis. Even now, we can restrict the water budget analysis to a single tree if desired and evaluate with some precision those transfers of water through and from the tree itself.

The ability to evaluate the different factors of the water budget over various time and space scales with a reasonable degree of accuracy permits increased understanding of important water relationships at a place or over an area. The use of simple temperature and precipitation data in a water budget analysis makes it possible to construct such a budget everywhere those data exist and over long time periods because historical climate records at many places go back more than 100 years. Thus, data on statistical frequencies or likelihoods of different water budget factors are available.

Origins of the Idea

Any effort to understand the development of the idea of water budget climatology must start with our understanding of the operation of the hydrologic cycle—the unending flow of water from ocean, to atmosphere, to land, and its return to the ocean in its solid, liquid, and gaseous states. Early theories of the operation of the hydrologic cycle go back at least to the Greek philosophers Plato and Aristotle, with their arguments about the source of water for stream flow (Biswas 1970). Neither Plato nor Aristotle thought that precipitation was sufficient to create the stream flows they observed, but they differed in their ideas about how water was moved to elevated land areas to create springs. Both agreed that a large subterranean cavern filled with sea water was the principal source for springs, but how this water moved upward from the subterranean cavity to emerge as fresh water from springs was a matter of disagreement between these scholars. The problem of the hydrologic cycle or the global water budget was finally clarified by several French naturalists (Bernard Palissy and later Claude Perrault and Edme Mariotte) and the English astronomer Edmond Halley in the sixteenth and seventeenth centuries. The French scientists proved that sufficient water fell as precipitation on a catchment to supply all observed flow in the stream, and Halley showed that evaporation from the oceans was sufficient to supply the measured precipitation.

There matters stood for nearly two hundred years, as scientists directed their attention toward obtaining quantitative observations of temperature and precipitation at an ever-growing (but unfortunately always changing) network of observing stations. People increasingly came to believe that type of vegetation cover or land surface treatment influenced evapotranspiration losses from the earth. The Timber Culture Act of 1873 had as one of its tenets the idea that planting more trees in the Great Plains of the United States would increase rainfall sufficiently to eliminate climatic hazards to agriculture (Thornthwaite 1937). The planting of shelterbelts of trees from North Dakota to Texas in the mid-1930s to eliminate the ravages of the existing Dust Bowl conditions in the Great Plains was still based in part on the idea that increased local evaporation would result in increased local precipitation, enabling us to pull ourselves out of the great drought.

C. Warren Thornthwaite, in his 1937 paper entitled "The Hydrologic Cycle Re-examined," showed that most of the moisture for precipitation over the continental United States came from warm maritime tropical, ocean-based air masses. The moisture that was evaporated into continental polar air masses moving southeastward across the

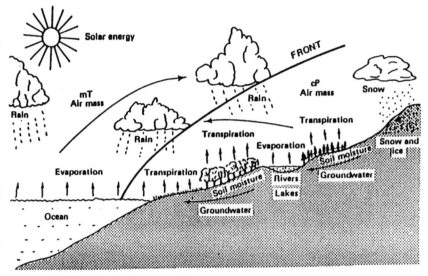

Figure 5.1. The hydrologic cycle. Source: van Hylckama 1956, with suggested modifications after Holzman 1937 and Thornthwaite 1937.

country would be held in these cool but always warming air masses (because warm saturated air will contain more water vapor than will cool saturated air). This suggested that one could not make the climate of the central United States more moist by planting trees, creating lakes, or building widespread irrigation enterprises in order to introduce more moisture into the local atmosphere.[1] Thornthwaite's inclusion of the differing moisture and temperature characteristics of mid-latitude air masses on the old hydrologic cycle diagrams rounded out our knowledge of the operation of the global water budget initially provided by Palissy, Perrault, and Mariotte (Figure 5.1).

Climatologists had long wrestled with the problem of determining the relative moistness or aridity of a place or region as they sought to define and delimit regional climates. Precipitation alone could not determine the degree of moistness of a climate for it was necessary to compare the supply of water from precipitation with the climatic demands for water—the evapotranspiration. Significant advances toward solving the question of the relative moistness of a climate were not possible until the late nineteenth century, when the network of temperature and precipitation observing stations became large enough to provide a reasonable understanding of the distribution of climatic conditions around the world. This work generally began under the leadership of the German meteorologist and plant physiologist

Wladimir Köppen. His first important paper on climatic classification appeared in 1900 and was based, in large part, on Auguste-Pyrame de Candolle's (1874) "physiologic" vegetation classification, which separated all plants into six major subdivisions, five that were related to particular ranges of mean annual temperature and one that included plants that had made adaptations to drought. Köppen thought that de Candolle's plant regions were also climatic regions, and so he merely attempted to find rational numerical values to define their boundaries. To separate desert from steppe climates, for example, he first placed the boundary where the rainiest month of the year had an average of only six days of rainfall (rainfall probability of 0.20). The division between forest and steppe was set where the rainfall probability in the wettest month was 0.36 (10.8 rainy days). Köppen did not identify a subhumid climate between the forest and the steppe (Thornthwaite 1943).

Köppen revised the formulation of his climatic classification many times during the next forty years, but he was never satisfied with his treatment of the moisture factor. He clearly articulated the problem when he pointed out that the roads in Russia (where he did much of his early work) were dry and dusty in the rainiest season of the year and wet and muddy in the driest season. By that he meant that precipitation alone could not adequately describe the availability of moisture. The problem that Köppen and all earlier climatologists faced in their struggle to develop a realistic climatic classification was how to express precipitation effectiveness or the relative moistness or aridity of a climate. But recognizing the problem was not equivalent to solving it and Köppen was never able to derive an adequate expression to represent the influence of the climatic demand for water on the available water supply.

The American climatologist C. Warren Thornthwaite received his doctorate in 1929 from the University of California at Berkeley under Carl Sauer. He began his teaching career at the University of Oklahoma in 1927 and soon became interested in problems of climatic classification and in ways to express the moisture factor in climate. Experiencing the beginning of those dry conditions later described as the Dust Bowl, he had first-hand knowledge of the significant role played by moisture in climate. Thornthwaite began his efforts to quantify precipitation effectiveness by collecting records of pan evaporation made at twenty-one stations in the western United States for periods ranging from four to twelve years, along with data on monthly precipitation and mean monthly temperature. Grouping these data by similar ratios of precipitation to evaporation, Thornthwaite was able to develop a precipitation effectiveness index (I) based on just monthly temperature and precipitation data. He used it as the basis for his 1931 paper on "The Climates of North America."

With continued work on evapotranspiration and the moisture factor in climate, over the following decade Thornthwaite began to understand that the actual evapotranspiration from a vegetation-covered area is distinctly different from the amount of moisture that can be lost by evapotranspiration if the vegetation is always well-supplied with water. Actual evapotranspiration depends not only on climatic factors but also on how much moisture is stored in the soil (with no soil moisture, there can be no evapotranspiration), on the type of vegetation (certain species are able to remove water from the soil more efficiently than others), and on land-management practices.

Consider, for example, the relative dryness of a desert area having 6 inches of annual precipitation. Actual evapotranspiration cannot exceed 6 inches because there is no other water to be lost. Energy from the sun in this desert area is sufficient, however, to evaporate 60 inches of water. Comparing precipitation with actual evapotranspiration results in a ratio of 6/6 or 1.0 while comparing precipitation with the climatic demand for water, based on the sun's energy, results in a ratio of 6/60 or 0.1, an index more expressive of the relative dryness of the desert area.

Thus, two aspects of evapotranspiration must be identified. Thornthwaite called these actual and potential evapotranspiration. He defined potential evapotranspiration for the first time in 1944 as "the water loss that will occur if at no time there is a deficiency of water in the soil for the use of vegetation" (Wilm and Thornthwaite 1944:687). Actual evapotranspiration can equal potential in rainy climates but will generally be less than potential in drier climates because the rainfall will not supply enough water to the soil for all the needs of the vegetation.

Developing an empirical expression for evaluating potential evapotranspiration from information on air temperature and day length, Thornthwaite checked its validity against available data. He measured potential evapotranspiration in moist, soil-filled lysimeters (or evapotranspirometers) exposed in fields having similar vegetation cover and moisture content. The results showed that potential evapotranspiration was difficult to measure, but that when it was done correctly, measured and computed values closely corresponded. As he wrote in his first paper on potential evapotranspiration,

An implied conclusion of this study is that [potential] evapotranspiration is independent of the character of the plant cover, of soil type, and of land utilization to the extent that it varies under ordinary conditions. This conclusion is contrary to currently accepted notions concerning evapotranspiration losses and I am reluctant to accept it myself; however, I have not yet found a reason for denying it. (Wilm and Thornthwaite 1944:689, 691)

Later checks on this conclusion were able to prove the validity of Thornthwaite's early feelings that potential evapotranspiration would not be influenced by type of vegetation, soil type, or method of land treatment. Potential evapotranspiration was a true climatic factor that could then be compared directly with the climatic factor of precipitation at a place on a daily, monthly, or annual basis in the form of a climatic water budget. The climatic water budget developed during the early 1940s was later modified by Thornthwaite and Mather (1955b) to allow for changes in the way moisture is removed from the soil by vegetation (actual evapotranspiration) and in the total amount of water that can be stored in the root zone of the soil.

Potential evapotranspiration (*PE*) and its use along with precipitation (*P*) in a climatic water budget became the basis for Thornthwaite's (1948) new classification of climate, in which monthly moisture demand (potential evapotranspiration) was compared with moisture supply (precipitation) to characterize the degree of moistness of the climate. This culminated in the statement of a new moisture index or ratio as follows:

$$I_m = \left(\frac{P}{PE} - 1 \right) 100.$$

At a place where *P* is greater than *PE*, the index, I_m is positive, where *P* is less than *PE*, I_m is negative, and where $P = PE$, I_m is zero. The moisture index was a rational indicator of the relative moistness or dryness of a climate. A recent modification of the moisture index expression by Willmott and Feddema (1991) provides an even more rational scale from $+100$ to -100, correcting a problem found in the earlier expression, that could go from -100 to infinity. The new expressions are

$$I_m = \left(\frac{P}{PE} - 1 \right) 100 \text{ when } P \text{ is less than or equal to } PE$$

and

$$I_m = \left(1 - \frac{PE}{P} \right) 100 \text{ when } P \text{ is equal to or is greater than } PE.$$

An Example of the Water Budget

Graphical or tabular comparisons of daily or monthly values of precipitation and potential evapotranspiration in the form of a climatic water budget are fundamental for understanding the nature, distribution, and extent of the water resources of a place or area. To illustrate water-budget bookkeeping, consider the average monthly march of water supply (precipitation) and climatic water need (potential evapotranspiration) at Wilmington, Delaware. The average monthly values are provided in Table 5.1 along with the remaining bookkeeping steps. Figure 5.2 shows these graphically.

At Wilmington, potential evapotranspiration, which is closely related to temperature, varies regularly through the year from low values of 0 millimeters in January and February when the mean temperature is near freezing to maximum values of 150 and 135 millimeters, respectively, in July and August. Precipitation, or water supply, is much less variable through the year, with each month receiving more than 75 millimeters. Low values are found in the fall and winter, with 78 millimeters in October and 80 millimeters in February; highest values come in the summer period with 119 millimeters in July and 128 millimeters in August. Thus the period with the greatest precipitation is also the period of greatest demand for water in terms of potential evapotranspiration. Because demand exceeds supply at that time, summer is also a period of dryness and need for irrigation.

Comparing precipitation with potential evapotranspiration on a monthly basis reveals that they never coincide. There is too much precipitation in fall, winter, and spring and too little in summer. In November, the precipitation over and above that needed for evapotranspiration is stored in the soil and results in the upper layers being brought to their maximum water-holding capacity. After the soil is holding all the water it can, any precipitation not needed for evapotranspiration goes to surplus (percolation to the water table and then to stream flow).

Precipitation is greater than the potential climatic water need from October through May on the average. The soil remains full of water, and some water is added each month to surplus. June is the first month in which the rapidly rising climatic water needs finally exceed the supply of water from precipitation. Some of this need is supplied by the water stored in the upper layer of the soil, but some is not supplied by either precipitation or stored soil water. This is, therefore, the water deficit. As the soil dries, water is less available from the upper soil layers and so plants obtain less and less of their needs from stored soil moisture.

Table 5.1. Factors of the Average Climatic Water Budget, Wilmington, Delaware

	J	F	M	A	M	J	J	A	S	O	N	D	YEAR
Temperature, °C	0.5	0.7	5.5	11.3	17.1	21.8	24.2	23.4	20.0	13.8	7.5	1.7	12.3
Potential evapotranspiration (PE in millimeters)	0	0	15	43	89	128	150	135	94	52	20	2	728
Precipitation (P in millimeters)	87	80	96	91	92	98	119	128	93	78	82	86	1130
P-PE	87	80	81	48	3	-30	-31	-7	-1	26	62	84	
Storage	150	150	150	150	150	122	99	94	93	119	150	150	
Change in soil moisture storage	0	0	0	0	0	-28	-23	-5	-1	+26	+31	0	
Actual evapotranspiration	0	0	15	43	89	126	142	133	94	52	20	2	716
Deficit	0	0	0	0	0	2	8	2	0	0	0	0	12
Surplus	87	80	81	48	3	0	0	0	0	0	31	84	414
Runoff	42	50	56	54	44	35	28	23	18	15	18	31	414
Measured runoff Shellpot Creek	47	50	64	51	39	22	24	20	15	11	33	39	415

NOTES:

Storage: amount of moisture stored in the root zone of the soil (assumed to have a maximum capacity of 150 mm depth).

Actual evapotranspiration: assumed to be equal to potential when P is greater than PE and equal to P + the absolute value of Change st when P is less than PE.

Deficit: potential evapotranspiration minus actual evapotranspiration.

Surplus: precipitation minus potential evapotranspiration when storage is at maximum storage capacity.

Runoff: assumed to be 20 percent of total surplus in any month, with the remaining 80 percent added to the surplus of the following month.

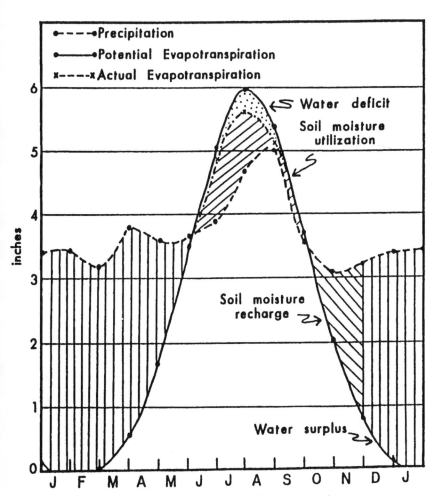

Figure 5.2. Average climatic water budget, Wilmington, Delaware.

The actual evapotranspiration (AE) or actual water loss equals the potential water loss during those periods when precipitation is greater than potential evapotranspiration. Sufficient water for evapotranspiration exists in those months. When precipitation is less than potential evapotranspiration, actual water loss equals the sum of precipitation and the water removed from the soil.

The surplus water, that moisture over and above what is needed for evapotranspiration or storage in the root zone of the soil, is ultimately lost as runoff. This does not occur immediately for it takes some time for the water to move down to the groundwater table and through the soil to emerge again in surface streams or rivers. The rate of removal by runoff, here assumed to be 20 percent of the available surplus in any month with the remaining 80 percent of the surplus held over and added to the available surplus of the following month, depends on the size of the watershed, on its slope and ground cover, and on the type of soil through which the water must move.

The last line in Table 5.1 provides information on the monthly measured stream flow in Shellpot Creek, which flows through northern Wilmington. As can be seen, annual agreement between measured and computed stream flow (or runoff) is quite close. Local watershed factors, which cannot be considered in these particular water budget computations, often lead to a lack of agreement on a monthly basis. The agreement between measured and computed runoff values is good verification of the whole computational model, although other comparisons (for example, computed versus measured soil moisture storage) are possible. The validity of the water budget, especially in middle-latitude areas without marked seasonal contrasts in moisture conditions, has been repeatedly substantiated in studies by myself and others (see, for example, Mather 1978).

Problem Solving with Water Budget Climatology

Factors derived from the evaluation of a climatic water budget—actual evapotranspiration, water surplus, water deficit, runoff—all provide useful, quantitative information on many aspects of the water relations at a place. One of the factors derived from the bookkeeping procedure is the actual evapotranspiration or the actual loss of water from the plant and soil surfaces, which, in almost all cases, is different from the potential water need. Measurement of actual evapotranspiration is difficult because this quantity depends on such factors as soil type and method of land cultivation, type of plant cover, and moisture condition of the soil profile. The ratio of the actual to the potential evapotranspiration appears to be closely related to the yield of crops or aboveground productivity of trees; the difference between the potential and the actual evapotranspiration provides a measure of the moisture deficit of a place, the amount by which the available moisture fails to satisfy the demand for water. Both the timing and the value of the moisture deficit are basic to any understanding of the economic potential of an area.

Thornthwaite employed this concept of moisture deficit in the development of his four-fold definition of drought. Regions where moisture demand exceeds moisture supply in each month may be classified as areas of permanent drought—desert areas where agriculture without irrigation is not possible. Regions where there is a marked seasonal variation in moisture, with precipitation exceeding climatic water demands in one season and climatic demands exceeding precipitation supply in another, are identified as having seasonal drought—a regularly recurring wet and dry climatic situation. Agriculture can be confined to the wet season or practiced in the dry season with irrigation. In moist areas, where some months experience more precipitation than climatic demand and other months experience the reverse (the shifts being contingent on the particular values of precipitation and climatic demand in any month), no regular seasonal pattern of drought exists. Thornthwaite identified these as areas of contingent drought—fairly unpredictable because they depend on the actual distribution of the factors of supply and demand in any month or season.

Finally, Thornthwaite identified a fourth type of drought—called hidden or invisible drought—in which precipitation always seems adequate and crop yields appear about normal. Invisible drought occurs when fairly frequent rains lull the farmer into thinking that conditions are satisfactory although demands may actually be slightly greater than supply. Crop yields suffer because the plants cannot obtain all the water they need, even in a fairly humid area. Invisible drought can only be eliminated by careful irrigation with the proper amount of water at those times when need exceeds supply. Determination of the changes in soil moisture storage on a daily basis gives information on the state of the moisture in the ground for use in the definition of drought and the scheduling of the time and amount of supplemental irrigation (Thornthwaite and Mather 1955a).

Information on the water surplus, the amount by which the precipitation exceeds the water needs when the soil is holding all the water it can, is fundamental in various hydrologic studies. By definition, the water surplus is the water that does not remain in the surface soil layers but is available for deep percolation to the water table and overland or subsurface flow to the water courses. Thus, information on water surplus, climatically determined from the water budget, provides knowledge of stream flow that can otherwise be obtained only from extensive stream-gauging installations.

One problem of immediate significance that can be solved by a water budget approach is determination of the leaching through solid waste landfills. Governmental regulations permit no leaching whatsoever, but in a climate in which precipitation exceeds actual evapotranspiration, water will collect in the landfill and ultimately overflow any plastic or clay containment structure beneath the landfill. To prevent this, some pumping of polluted leachate is necessary. Determining the optimum way to treat this polluted leachate requires knowledge of the volume available. Knowing the landfill's construction, the density of material in the landfill and the moisture content of the incoming landfill materials, one can then evaluate inflow by precipitation minus evaporation losses and overland runoff. The difference is the leachate volume once the landfill is saturated. Depending on this volume, plans can be made to ship the leachate to a water treatment plant, to build such a plant and treat the effluent on site, or to mix it with fresh water and to dispose of it by spray irrigation in nearby scrub woodlands.

In oceanographic work, the water surplus is important in determining the amount of freshwater flow from the land to the ocean. This is particularly significant in the case of bays, estuaries, and seas that are almost entirely landlocked, for the volume of freshwater runoff to the water body will be important in determining the salinity, density, and other characteristics of the water. For example, at the request of a petroleum company operating in Lake Maracaibo, Venezuela, Thornthwaite applied the climatic water budget to determine what would happen to the lake level if a dam were constructed across its narrow entrance so that its water volume would only be supplied by basin runoff (Carter 1955). The results showed that the lake would soon become a fresh water body and that current lake levels would be maintained. In contrast, applying the same analysis to the Mediterranean Sea, assuming a damming of the Strait of Gibraltar and the Suez Canal, established that the sea would evaporate to several small brackish lakes on either side of Malta (Carter 1956). The scheme to turn the sea into a freshwater lake to supply extensive irrigation along the north African coast would not be a feasible undertaking.[2]

The water budget provides information on the detention of moisture on and within the land areas of the earth through the year. Water budgets from approximately 15,000 stations around the world were used by Thornthwaite and his associates (van Hylckama 1956) to verify conclusions reached by oceanographers on the basis of sea level measurements that vast amounts of water move from ocean to land during winter and early spring and back again from land to ocean in fall. Thornthwaite also related the value of water storage in the ground to the ability of vehicles or humans to move over unpaved surfaces (soil tractionability) (Thornthwaite et al. 1958). On all but the most sandy soils, an increase in soil moisture content to or above its moisture-holding capacity will result in a loss of bearing capacity and shearing strength and an increase in stickiness—limiting the ability of humans and vehicles to move over unpaved surfaces. On sands, the reverse is true; as the moisture content increases even to the point where free water stands on the surface, tractionability conditions improve. Monthly maps of tractionability have been prepared for many parts of the globe at the request of the Department of Defense for strategic planning purposes, but the same information has great utility to farmers and construction companies planning farming, construction, or earthmoving operations.

Thornthwaite and his associates (Mather 1959) related annual values of the moisture index and potential evapotranspiration to the natural vegetation of an area. Those water budget factors express both heat and moisture factors of primary significance to plant growth and development. Thornthwaite also used the quantity potential evapotranspiration as the basis for his crop calendar, a device that changes the ordinary civil calendar into a plant development calendar. It expresses the length of days in terms of the available amount of energy for plant development on that day and so can be used to determine harvest dates once germination is known. Or, if desired harvest dates are specified, the crop calendar permits the scheduling of planting dates. When employed at Seabrook Farms in southern New Jersey, Thornthwaite's crop calendar prevented harvest and processing bottlenecks, smoothed the flow of vegetables to the processing plant, and reduced the need for costly overtime labor. Writing for *Fortune* magazine, Herbert Solow (1956) thought the application was so unusual that he described it as "operations research down on the corporate farm."

Probably more important than the use of the water budget in any one of these many lines of research is the fact that with the ability to obtain information on all phases of the moisture relationships of an area from readily available climatic data comes the ability to determine values of these moisture parameters over long periods of record. This provides data for periods when measurements were not made, and it also permits the accumulation of a record that can be used in statistical studies. Many problems, such as the determination of probabilities of the amount of surplus water available for stream flow, of the occurrence of soil moisture conditions affecting the movement of humans or vehicles over offroad surfaces, or of drought, can only be solved in this manner at present because measured values are not available for a long enough period. Such statistical studies provide the basis for forecasting the outcome of a program of action and evaluating its economic feasibility.

Significance of the Water Budget Concept

Over the years, people have undertaken a great many modifications of the surface of our earth. Nearly all such modifications have, in one way or another, influenced factors of the water and energy budgets of the earth's surface and atmosphere. Consider first the cutting of forests and the building of cities. These actions have led to a significant rearrangement of heat sources at the surface of the earth; they have also affected the surface roughness and, thus, the turbulent exchanges near the surface. Clearly, opportunities to exchange moisture with the atmosphere have also been influenced. In addition, we humans have dammed water in reservoirs and ponds, drained lakes, marshes, and low-lying areas, overpumped and thus lowered water tables, and undertaken artificial recharge to maintain subsurface water levels. We have changed river flows by engineering works; modified overland and subsurface water movement to streams by remolding soils, leveling land, and installing sewers and tile drains; and have created artificial rainfall conditions through the establishment of extensive irrigation programs. The classical relations among precipitation, evapotranspiration, and runoff have been disrupted by human interference with the natural operation of the hydrologic cycle.

The use of the climatic water budget not only permits evaluation of the water resource changes that have occurred as a result of human actions, but also gives us an understanding of what will result if different development scenarios are implemented in the future. Consider the problem of global warming through an increase in carbon dioxide (CO_2) and other trace gases in the atmosphere. Substitution of new values of temperature and precipitation obtained from global models using increased values of CO_2 permits evaluation of what might happen to stream flow, groundwater recharge, snow accumulation, lake levels, and other hydroclimatic factors under global warming conditions. This, in turn, permits adequate time for planning rational responses.

The flexibility of the climatic water budget is one of its greatest strengths. The model is so constructed that it is possible to begin with whatever temperature and precipitation data exist or are hypothesized for a particular place or area and then to build into the computational process a whole range of local environmental conditions to represent the particular situation to be evaluated. Type of vegetation, type of soil, rooting depth, soil moisture content, rate of removal of soil moisture by evapotranspiration, the use of irrigation, and the use of storm sewers or other devices to control overland runoff are just some of these factors. Quantitative results are made available for analysis or even for statistical manipulation if longer periods of record are considered. These various advantages make the use of the water budget approach invaluable in water resources studies.

We must understand the essential unity of water in all phases of the hydrologic cycle. It is not realistic to speak about changes in storage or runoff at a place without recognizing that those changes must be balanced by corresponding changes in other aspects of the cycle locally, regionally, or even globally. Because water decisions in one area affect the availability of water supplies and the quality of those supplies in other areas, water budget analyses will play increasingly important roles in studies of resource development not only of small local areas but also of larger, more complex regions. The current widespread use of water budgets in the solution of so many different types of problems testifies to the utility of the approach. From such understanding comes an increased ability to plan rationally and to use limited resources more wisely. Each computation and application of the climatic water budget to solve a particular hydroclimatic problem not only reconfirms the reputation of Charles Warren Thornthwaite, its originator, as the most ingenious and forward-thinking climatologist of the twentieth century, but also reestablishes the climatic water budget as one of the most significant concepts in geography of this century.

Notes

1. Under certain circumstances, such as in the Amazon Basin, where only a single warm moist air mass is involved, local rainfall may be appreciably augmented by local evapotranspiration. Deforestation of the Amazon Basin, thus, could result in a significant decrease in precipitation.

2. These differing results are a function of the amount of runoff from the surrounding land areas. In the Venezuela example, $P - PE$ from the basin area surrounding the lake was greater than evaporation losses from the lake in all but two months of the year. The lake would remain full of water even if the entrance was dammed. In the Mediterranean case, the small amount of runoff from the rivers in Europe (very little water ran off from the Middle Eastern and North African portion of the basin) was less than the evaporation from the sea surface. The Mediterranean would slowly lose water until a new equilibrium between runoff inflow and evaporation loss was established with the creation of several small saline lakes.

References

Biswas, A. K. 1970. *History of Hydrology*. Amsterdam: North Holland Publishing.

Carter, D. B. 1955. The Water Balance of the Lake Maracaibo Basin. *Publications in Climatology Laboratory of Climatology* 8(3):209–227.

———. 1956. The Water Balance of the Mediterranean and Black Seas. *Publications in Climatology Laboratory of Climatology* 9(3):125–174.

de Candolle, A. 1874. Constitution dans le Règne Végétal de Groupes Physiologiques Applicables à la Géographie Botanique Ancienne et Moderne. *Bibliothèque Universelle, Archives des Sci. Phys. et Nat.* 50(n.s.):5–42.

Holzman, B. 1937. Sources of Moisture for Precipitation in the United States. *U.S. Department of Agriculture, Technical Bulletin* 589:1–41.

Köppen, W. 1900. Versuch einer Klassifikation der Klimate, Vorzergsweise Nach Ihren Beziehungen Zur Pflanzenwelt. *Geograph. Zeitschrift* 6:593–611, 657–679.

Mather, J. R. 1959. The Moisture Balance in Grassland Climatology. In *Grassland*, ed. H. B. Sprague, 251–261. Washington, D.C.: American Association for the Advancement of Science.

———. 1974. *Climatology: Fundamentals and Applications*. New York: McGraw-Hill.

———. 1978. *The Climatic Water Budget in Environmental Analysis*. Lexington, Mass.: Lexington Books.

Solow, H. 1956. Operations Research Is in Business. *Fortune* February:148–149.

Thornthwaite, C. W. 1931. The Climates of North America According to a New Classification. *Geographical Review* 21(4):633–655.

———. 1937. The Hydrologic Cycle Re-examined. *Soil Conservation* 3(4):2–8.

———. 1943. Problems in the Classification of Climates. *Geographical Review* 33(2):233–255.

———. 1948. An Approach Toward a Rational Classification of Climate. *Geographical Review* 38(1):55–94.

Thornthwaite, C. W., and J. R. Mather. 1955a. Climatology and Irrigation Scheduling. *Weekly Weather and Crop Bulletin* National Summary of June 27.

———. 1955b. The Water Balance. *Publications in Climatology Laboratory of Climatology* 8(1):1–104.

Thornthwaite, C. W., J. R. Mather, D. B. Carter, and C. E. Molineux. 1958. Estimating Soil Moisture and Tractionability Conditions for Strategic Planning. *Air Force Surveys in Geophysics* 94, AFCRC-TN-58-202, Geophysics Research Directorate 1–56.

Van Hylckama, T.E.A. 1956. The Water Balance of the Earth. *Publications in Climatology Laboratory of Climatology* 9(2):59–117.

Willmott, C. J., and J. J. Feddema. 1991. A More Rational Climatic Moisture Index. *Professional Geographer* 44(1):84–87.

Wilm, H. G., and C. W. Thornthwaite. 1944. Report of the Committee on Transpiration and Evaporation, 1943–44. *Transactions, American Geophysical Union*, pt. 5:686–693.

6

Human Transformation
of the Earth

William B. Meyer and B. L. Turner, II

In the phrase "human transformation of the earth," the adjective is not superfluous. For billions of years, the earth has been transforming itself, changing both ephemerally and irreversibly at scales from the local to the global. The recognition of natural environmental transformation is an idea with a long and interesting history of its own. Recognizing such change has often gone along with the idea that we should do what we can to stop it. The first assessment of what global warming meant for humankind was an entirely positive one; it offered a means "efficaciously to regulate the future climate of the earth and consequently prevent the arrival of a new Ice Age" (Ekholm 1901:61). In 1938 another pioneering discussion noted that if human actions induced a global warming, not only would agriculture benefit, but "the return of the deadly glaciers should be delayed indefinitely" (Callendar 1938: 236). At that time, the return of the glaciers was considered to be the main threat to sustained human occupation of the earth, just as storms, floods, and droughts overshadowed most human-induced environmental problems. But as the twentieth century draws to a close, human action has become unmistakably the principal force altering the earth's surface, and in the minds of many, it has become far more threatening than any natural force of change.

The American geographer, lawyer, public servant, and philologist George Perkins Marsh (1801–1882) was not the first person to notice the human transformation of the earth. The creation of what Cicero called "a second nature" has been a matter of comment since classical times and no doubt much longer (Glacken 1967). Nor was Marsh the first to suggest that much of the change brought about by humankind was for the worse. Such suggestions, too, "extend back about as far as one is willing to look" (Clark 1986:8). Most ideas are ancient; Darwin was not the first to discuss evolution by natural selection, Marx class struggle, Freud the unconscious, or Adam Smith the hidden hand. Yet often for good reason, an idea is tagged with the name of an individual who was not the first to enunciate it in some form. Originality at this level is never absolute, but lines can be drawn—if sometimes with much difficulty—between those who did and did not "pursu[e] an idea or finding seriously enough to make its implications evident" or state it "definitely and emphatically enough so that it cannot be overlooked" (Merton 1967:14, 16). If George Perkins Marsh cannot quite be called the author of the idea of the human transformation of the earth, the idea is nonetheless, and rightly, linked more closely to his name than to anyone else's.

In his book *Man and Nature; or, Physical Geography as Modified by Human Action* (published in 1864), Marsh showed as no one had done before how much humankind had altered the earth and why it mattered. He documented the extent and profundity of the human impact. He added up the damage as well as the good that had been done. Most importantly, he showed how human actions produced changes that far exceeded in degree and variety the intended results. Though deeply indebted to his predecessors, Marsh transcended them, placing "for the first time . . . the results of their investigations where they belonged—in the forefront of human history" (Glacken 1956: 83). It was a novel point of view in his day and has become increasingly common since. Today the forefront is widely acknowledged as the place where the topic belongs.

Marsh was not a geographer in the modern sense of being a professional member of an academic discipline. He shared this amateur status with many of his distinguished contemporaries, now claimed by such disciplines as founding ancestors—Charles Darwin (biology), Auguste Comte (sociology), and Lewis Henry Morgan (anthropology) among them. The only courses Marsh taught, at Columbia University (1858–1859) and the Lowell Institute (1860–61), dealt with the history of the English language. His diversity of interests enriched his geography. *Man and Nature* drew on personal observation of landscape change in the United States, Europe, and the Middle East (Lowenthal 1958). Marsh's command of languages opened many libraries of published foreign material to him, and his varied practical experiences kept his writing clear and his interpretations down to earth.

Marsh in any case identified himself as a geographer, if not only as one. *Man and Nature* remains a distinctively geographical classic. The matter with which it deals is by tradition more geography's domain than any other discipline's; by some of the most enduring definitions of the field it is also geography's: the definition revived and expounded by Tuan (1991), for example, "the study of the earth as the home of people." *Man and Nature* was explicitly an inquiry into how the tenants had been treating that home. Its verdict was not a kind or reassuring one. "[W]e are," Marsh (1965:52) said, "even now, breaking up the floor and wainscoting and window frames of our dwelling, for fuel to warm our bodies and seethe our pottage."

The central chapters of the volume dealt with human effects on the plants and animals, the forests, the waters, and the sands of the globe. In each case Marsh traced the further repercussions of change into other realms of the environment: the climatic consequences of forest removal, for example. An opening chapter explored the topic in a more general vein. A conclusion addressed the possible consequences of a number of large projects of landscape transformation then being projected. Throughout, Marsh presented as much evidence bearing on the topic at hand as was available to him. He drew on personal experience and on the writings of scientists, engineers, historians, travelers, even poets. Throughout, he assessed their claims critically instead of taking them at face value.

Marsh's importance lay less in recognizing an earth transformed than in defining its transformation as a problem. Defenders of the idea of progress had long found their surest support in human-induced environmental change. Widely taken for granted in Marsh's lifetime and later was the idea that we have made the planet a far better place than it once was. In antebellum America, "the idea of progress was the most popular American philosophy, thoroughly congenial to the ideas and interests of the age," and "the underlying note in much of the American concept was the secure feeling derived from an observation of man's conquest and control of the immense forces of nature" (Ekirch 1944:267). Progress and environmental change were so closely linked in the nineteenth-century Western mind that they hardly seemed separate ideas. One of Marsh's principal achievements in the pages of *Man and Nature* was to pry them apart.

He did not, however, manage to separate them in the minds of his contemporaries. American and European readers greeted Marsh's volume enthusiastically. Yet its deeper lessons "were submerged in the tide of opinion which saw progress everywhere in the beneficent command which man had attained over nature" (Glacken 1956:83). Western prophecies of the future of the earth continued to see it becoming ever more smoothly subservient to its human masters. Practice was even less affected than principle.

In one approach to the history of ideas, "the writers of the past are simply praised or blamed according to how far they may seem to have aspired to the condition of being ourselves" (Dunn 1980:19n); the highest accolades are reserved for those who appear to have believed what we believe and to have shared the greatest number of our own prejudices and presuppositions. It is an easy approach to adopt but a good one to avoid. To describe Marsh as having been "ahead of his time" would do neither him nor his contemporaries justice. It obscures both their reasons for believing what they did and Marsh's for sometimes believing otherwise. Today we often suppose that people in the nineteenth century freely drained wetlands because they mistakenly viewed them as useless and repulsive, whereas we have begun to preserve them because we know better. Yet the benefits of drainage were real and substantial: the creation of urban or farm land, the improvement—by the tastes of the day—of the scenery, and the reduction of malaria and yellow fever, the insect carriers of which bred in stagnant water. Converting wetlands (Figure 6.1) signaled no disdain for future generations, just the opposite. The land was being improved, at great labor and expense, for posterity's as well as for the present's sake. Like anything else, wetlands and wilderness have become valuable only as they have become scarce, and also as they have become less dangerous.

Marsh, too, applauded when "worthless and even pestilential land . . . has been rendered both productive and salubrious" by drainage (Marsh 1965:284). A presentist reader would subtract points from his grade for this dismaying lapse from today's wisdom. Yet Marsh upheld the consensus of his time not by dogmatically considering any human-engineered change an improvement, but by weighing what was known about the effects of drainage. He paid more attention than most of his contemporaries did to the costs of the operation. As he observed, draining land lowered springs and wells; it heightened the variations of stream flow, contributing to damaging floods and to seasons of low water; and it might adversely affect local climate and wildlife. Here the gains seemed to outweigh the losses; in many other cases, Marsh found the long-term costs far heavier.

A balanced view is open to caricature from both sides. It was greatly to Marsh's credit that, having detached the idea of environmental transformation from an easy assumption of progress, he did not simply reattach it to one of decline or decay. Yet in his own day and in subsequent decades he was apt to be accused of exaggerating damage.

Figure 6.1. Perceived improvement of wetland by drainage, early twentieth century: (top) before, (bottom) after. Source: Hodge and Dawson (1918:134).

A half-century after *Man and Nature,* a British geographer reviewing the same field spoke patronizingly of Marsh as "a pessimist, evidently," too inclined to dwell on the dark side, when, in fact, damage was becoming a thing of the barbarous past and "under an all-wise Providence [the earth] is being subdued" (Lucas 1912:452, 453). In the early through the mid-twentieth century, progress through massive environmental transformation represented a ruling assumption in Western countries and official dogma in Communist states, reaching its disastrous apogee in Stalin's Soviet Union and Mao's China (Weiner 1988; Smil 1984).

From its high-water mark the idea of progress through environmental change has receded fast. Increasingly, predictions of the future are not of progress induced by environmental transformation but of decline flowing from the same source, of an earth impoverished and poisoned by its human inhabitants. But it does little good to replace one monolithic schema with another. As we have learned that progress can occur but cannot be taken for granted (Nisbet 1980), an earth transformed can be, but by no means necessarily is, either an earth damaged or an earth improved.

Man and Nature can be read today with profit for many lessons about the human relationship to the physical environment, for "insights still unsurpassed" (Lowenthal 1990:133). It cannot, of course, be read as it was in 1864: as an up-to-date compendium of the best knowledge available on the topic. A picture of the recent history and present state of the biosphere is available in *The Earth as Transformed by Human Action* (Turner et al. 1990). The symposium on which it was based attempted to sum up our knowledge of net human impacts on the various realms of the biosphere. The results offer a chance to assess the accuracy of the idea. How and how much has the earth been transformed by human action?

Forest loss, Marsh's greatest concern, has accelerated. As large an expanse of forest has vanished since the publication of *Man and Nature* as was cleared in the span of postglacial human history before it (Table 6.1). Other changes that Marsh discussed have likewise continued at escalating rates: cropland expansion; the extinction, thinning, and transplantation of species; and soil erosion and degradation. Annual human use of water has grown thirty-five-fold in the past three centuries. It now takes up a large fraction of the yearly flow of the hydrologic cycle. And since 1864, known human-induced environmental changes have begun to include many that can be detected only with highly sensitive instruments. Not only the faces of the earth but the invisible flows of material and energy through them have been transformed.

Table 6.1. Global Deforestation: Estimated Areas Cleared
(in thousands of square kilometers)

Region	pre-1650	1650–1749	1750–1849
North America	6	80	380
Central America	12–18	30	40
Latin America	12–18	100	170
Oceania	2–6	4–6	6
Former USSR	42–70	130–180	250–270
Europe	176–204	54–66	146–186
Asia	640–974	176–216	596–606
Africa	96–226	24–80	16–42
Total highest	1,522	758	1,680
Total lowest	986	598	1,592

Region	1850–1978	Total: high estimate	Total: low estimate
North America	641	1,107	1,107
Central America	200	288	282
Latin America	637	925	919
Oceania	362	380	374
Former USSR	575	1,095	997
Europe	81	497	497
Asia	1,220	3,006	2,642
Africa	469	759	631
Total highest	4,185	8,057	
Total lowest	4,185		7,449

SOURCE: Williams (1990:180).

Inputs from industry and land use have considerably increased the flows of the major biogeochemical cycles—carbon, sulfur, nitrogen, and phosphorus. Human activities release many metals, including ones as toxic as mercury, cadmium, and lead, into the environment in far greater quantities than does natural weathering. Synthetic substances unknown in nature have been created and released. The full effects on ecosystems, climates, and human health in most cases can only be guessed. The changes that have already occurred are dwarfed by those that are likely to occur as world population grows to or beyond ten billion and as higher standards of living are everywhere sought.

We know much more than did Marsh and his contemporaries about the depth and antiquity of human impact. Landscapes once thought to be pristine turn out to have long histories of human occupance and reshaping. The face of the western hemisphere was deeply marked and in some places scarred by its human occupants before the arrival of European settlement (Butzer 1992; Denevan 1992; Turner and Butzer 1992). On the other hand, Marsh overstated the human role in some phenomena—deserts, floods, and climatic change, for example. He tended to interpret nature as stable and any rapid change as necessarily human in origin. A much better knowledge of the natural dynamics of change has made it clear that an earth without human occupants would still not be a stable one. A problem evident in *Man and Nature*—the Eurocentrism of the available literature—has persisted. We still know much more about changes in certain parts of the world than in others, and the better-known areas are not necessarily the most significant; East Asia's environmental past and present, for example, remain woefully underrepresented in the Western literature.

What has changed most since 1864 is the attention now given to the human dimensions of environmental change. Marsh did not ignore, but neither did he systematically address, the social causes, contexts, and consequences of environmental transformation. The major schools of nature-society research in modern American geography have shown a steadily increasing engagement with such questions. The pursuit of answers has carried researchers across the boundaries of many neighboring disciplines, and indeed sometimes far out of sight of the environmental changes themselves.

As described in Chapter 4 of this book, the "Chicago School" of Gilbert White and colleagues began by asking why narrow technical approaches to environmental hazards did not account for observed results. Why did rising expenditures on flood control and prevention works not lower flood losses? Why did people settle in clearly dangerous environments (Burton, Kates, and White 1978)? The search for understanding has taken risk-hazards researchers into the study of individual perception and decision making, of social structures and constraints, and of technological as well as environmental hazards (Hewitt 1983; Kates, Hohenemser, and Kasperson 1985).

The "Berkeley School" of Carl O. Sauer (1889–1975) is the source of the second major tradition of nature-society research in geography.

Sauer co-chaired a major 1955 symposium dedicated to Marsh and dealing with *Man's Role in Changing the Face of the Earth* (Thomas 1956). His work and that of his students initially focused on visible changes in the land, largely in rural Latin America and other peasant societies, more than it did on the social processes driving those changes. Cultural and political ecology, Berkeley's heirs in American geography today, have attempted to fill that gap (Price and Lewis 1993). The shift in focus in the environmental movement generally from "nature preservation" to "sustainable development" has taken the same direction from biophysical changes to their social roots and consequences.

In short, *Man and Nature* invites much correction in detail and expansion in scope. Even the *Earth Transformed* volume already requires some (Turner, Kates, and Meyer 1994). Yet many of the principles Marsh enunciated retain their validity. A closer look at those principles illustrates this point and another as well: that, as Lovejoy (1944) noted, an idea can be used in many conflicting ways depending on the other ideas with which it is combined.

If ideas have opposites, that of the human transformation of the environment would surely seem to be that of what geographers call, usually in derision, environmental influences or environmental determinism. What notions could be less similar than that of humankind shaping the earth and the earth shaping humankind? From a different angle, though, the two ideas look complementary rather than conflicting. For if the environment does not much affect our affairs, why should we worry about how we are affecting it? Marsh noted that until earlier geographers had demonstrated the effects of physical features on human life, there had been no reason to study, as he did, the relation as it ran in the opposite direction (Marsh 1965:14). Close attention to the earth transformed has often coexisted with a fierce denigration of environmental determinism; there is no necessary relation between belief or disbelief in the two views. Nor, as Bassin (1992) has established, is there, as was once assumed, any necessary connection between environmental determinism and political projects such as imperial expansion.

The earth transformed is an idea as promiscuous as environmental determinism in its political ties. As already noted, it long flourished and still widely exists in the form of celebrating the earth improved. Concerns over the earth damaged can be roughly divided into those of conservationism, preservationism, and environmentalism. The three perspectives, to simplify an already simplified typology, see the earth as a storehouse of natural resources, as a coexisting realm having rights at least equal to humankind's, and as a complex life-support system with which we tinker at our peril. Hence their differing concerns over the earth transformed. Often united, the three perspectives have occasionally clashed. Even within each perspective, the same facts and concerns have informed radically different prescriptions. A central belief of modern environmentalism is that human activities are fraying the fabric of the earth in a way that threatens its continued habitability. The same evidence that other environmentalists cite to argue that something drastic must be done, many Christian fundamentalists seize upon as proof that nothing can be done and that the end is near (Boyer 1992:331–337; Curry-Roper 1990).

It is no coincidence that Marsh's stature has been most firmly established in the era—the post–World War II decades—that has witnessed the growth of environmentalism. In its emphasis on the unintended and diverse consequences of human interference in a web of ecological complexities, *Man and Nature* is closer to the environmentalist perspective than to the conservationist or preservationist. "So far as they are purely the calculated and desired results" of human actions, Marsh wrote, changes in the land were simple and straightforward matters. What interested him were those produced "as unforeseen though natural consequences of acts performed for narrower and more immediate ends" (Marsh 1965:19). As he argued out with a wealth of examples, large changes could be the result of nobody's design. Indeed, changes could be brought about that were precisely the opposite of the ones intended. Farmers killed birds that ate their crops, only to find their crops devastated by insects that the birds had previously kept in check. Such changes—"the collateral and unsought consequences of human actions"—Marsh (1965:456) thought far more common and more important than the "direct and desired results." In altering the environment, he contended, we do not know our own strength; we often fail to produce the results that we seek, but we are certain to produce others.

Most contemporary global environmental change is a matter of unintended impacts. Though some species are deliberately exterminated, many more are reduced in numbers inadvertently through overhunting or habitat destruction. Fossil fuels are not burned in order to raise the carbon dioxide content of the atmosphere or to acidify rain, though those may be the consequences. What the eighteenth-century Scottish philosopher Adam Ferguson said of history holds true for environmental history and especially for that of the twentieth century: that it "is the result of human action, not of human design" (quoted by Elster 1989:91). The depletion of the stratospheric ozone layer is the most striking example to come to light in recent years of the gap between intent and outcome. When chlorofluorocarbons (CFCs), highly stable synthetic chemicals, were developed for use as refrigerants and propellants, nobody foresaw or could have foreseen their distant, unexpected, and long-lasting impacts (Stern, Young, and Druckman 1992:54–60). Anyone assigned the role of devil's advocate at Marsh's canonization could point out that many of his own examples of distant and unintended effects are no longer accepted as valid and that the era in which he lived was one in which actions and effects were generally straightforwardly connected (Hägerstrand and Lohm 1990:621); the best examples of large and distant unintended impacts are from the twentieth century. Humankind has acted since 1864 as if trying to prove Marsh right.

The unintended consequences of human action have thus kept the central place in environmental studies that Marsh gave them, although they have become a matter of increasing interest to the social sciences as well (Boudon 1982). During the 1960s and 1970s environmentalists emphasized to great effect the complexity of the natural world and the risk of unforeseen disasters when humankind meddled with it. At the same time, some social scientists were updating classic arguments that, because of the complexity of society, interventions meant to improve conditions often made them worse. Environmentalists, for example, presented new versions, with chemical pesticides as the agent, of Marsh's cautionary tale about how killing marauding birds in order to protect crops had the opposite of the desired effect. Economists argued that price controls, meant to maintain the supply of affordable commodities, ended up reducing it by removing incentives for their supply. The same morals have been drawn in each case: that simple attempts to "engineer" improvements are naive and shortsighted, that good intentions do not guarantee good results, and that interference should be avoided when possible and should be modest and restrained when undertaken.

Paradoxically, these identical lessons have been drawn on opposite sides of the political spectrum. It has generally been argued, at least in the twentieth-century United States, on the political left that nature, and on the political right that society or the market, knows best and should not be interfered with. Arguments in the social sphere about unintended consequences have been "most typical of conservative attacks on existing or proposed progressive policies, and their major proponents have been conservative thinkers" (Hirschman 1991:7); just the opposite has been typical in the environmental sphere. Yet even these clashing affinities lie in the uses made of the idea of unintended consequences and not in the idea itself. For one thing, the idea that outcomes were indeed unintended offers those who caused harm a convenient way of evading responsibility for it. For another, as Hirschman (1991) notes, the thesis that intervention typically fails to solve the problem at hand and may make it or others worse has a counterthesis that also relies on the notion of unintended consequences: interventions spin off surprise benefits. What is unforeseen need not be undesirable.

These theses and countertheses are useful tools for thought that cease to be useful when they become tools for avoiding thought. The fact that actions produce unforeseen results does not mean that action should be avoided. It does suggest trying to foresee consequences better, identifying and repairing institutions that perversely reward actors for ignoring the damage they cause, and acting with more caution.

The idea of the earth transformed by human actions has unquestionably become widespread. It is not difficult today to envision it giving rise to a new set of professional subfields and disciplines for the next millennium. We now speak of "global change science" comprised of physical (earth system science) and human dimensions (ESSC 1988; Stern, Young, and Druckman 1992). The growth of professional concern is paralleled by the "greening" of political and economic institutions, international relations, and public opinion. Global-level negotiations discuss ways to control climate change, ozone depletion, and biodiversity loss, and major political figures write on the subject of global change (see, for example, the 1987 "Bruntland report" prepared by the WCED—the World Commission on Environment and Development—and Albert Gore's 1992 book, *Earth in the Balance*). Studies find environmental concern strong, persistent, and spreading in the public around the world (Mitchell 1989; Inglehart 1990; Dunlap, Gallup, and Gallup 1993). Educational institutions rush to compose appropriate curricula (Blackburn 1993).

The eminently geographic idea of the earth transformed now looms large indeed. Why? The simplest answer would be that public concern is the result of enlightenment. Scientific advances have been assimilated; the evidence has forced the public to conclude that environmental transformation is a major problem. That is the simplest answer. There is much to it, but as an explanation it is too simple. Understanding of environmental issues is poor even among those professing environmental concern (Arcury and Johnson 1987; Arcury and Christianson 1993).

Polls and surveys find the American public well acquainted with the term "global warming" and concerned about the phenomenon it denotes. Yet more probing studies suggest that the public, by and large, may thoroughly misunderstand every important aspect of it—its causes and mechanisms, its likely consequences, and the measures that might be taken in response (Kempton 1991). If such is the case, scientists and the public can hardly even be said to be worrying about the same thing—though they may think they are. Among New York State high school students, "awareness" and concern about environmental issues substantially outrun understanding of them (Hausbeck, Milbrath, and Enright 1992), making it hard to argue that understanding drives awareness and concern. A wide gap separates American expert and public beliefs about the relative seriousness of various environmental risks. National policy, until recently, followed public concerns more closely than expert ones, biasing the allocation of resources and regulatory effort away from the ways in which they could most efficiently have reduced losses in life, health, and property (U.S. EPA 1987, 1990).

> Among some fifty college students whom I once happened to ask (in planned illustration of the incredible ignorance concerning natural objects that characterizes young Americans of today) the name of the tree, an American elm, that they could see through the classroom window, none was able to identify it; some hesitantly suggested it might be an oak, others were silent; one . . . said she guessed it was just a shade tree. (Nabokov 1964, vol. 3:9)

That was decades ago. It is safe to say that matters are no better today. They are probably worse. If American college students are blind to the simplest aspects of biodiversity in their own backyards, why does one find them convinced that loss of biodiversity in Amazonia—a region most of them would be hard-pressed to locate on a world map—is a terribly important problem? The "hamburger connection" has become a part of American cultural literacy. One routinely hears references to the destruction of Amazonian rainforest for pasture land to meet Northern demand for cheap beef. In this case, alas, cultural literacy is geographic ignorance. If the number protesting the connection is small compared to the number still served by fast food restaurants, the number of protestors who understand it is smaller still, who realize that the connection involves Central America, and not that convenient automatic category "the Amazonian rainforest," that it involves tropical dry or deciduous forest, not rainforest, and, early claims nothwithstanding, that beef production for fast-food consumption in the United States is a minor factor compared to other forces of change in Central America (Browder 1988; Hecht and Cockburn 1989; Parsons 1988)? The idea has spread in part because it so vividly, though erroneously, connects everyday actions with distant consequences. But it is hard not to suspect that the idea's popularity among educated and affluent Americans also has something to do with the opportunity it gives them to assert their superiority to other members of their society.

Clearly more is going on than the simple lay assimilation of scientific truths. We may be convinced that we understand the problems of global environmental change when all that we really understand is that concern over it seems to be typical of people we do or do not like. We may think that we are taking a stand when we are only striking a pose. Nor is it clear that we can often avoid doing so. Nobody, after all, can spare the time to become an expert on every environmental problem and judge its seriousness without relying on others. How surprising is it, then, that many other factors—trust, affinity, and affection—enter into the decision about whom and what to believe?

Environmental risks exist whether we notice them or not, but the grounds on which we choose which ones to worry about have much to do with who we are as well as with what they are. As anthropologists and sociologists emphasize, beliefs about nature are not unrelated to beliefs about, and conflicts within, society. "Tropical deforestation . . . is a cause which governments and the public in countries which have no tropical forests have found it easy to espouse" (Brookfield 1992:95). The countries of the developing tropics are understandably suspicious of global environmental agendas drafted for them by the developed world in which deforestation's contributions to global warming are emphasized as the problem and population control and limits on economic growth as the answers. Ideologues from laissez-faire libertarians to authoritarian collectivists use environmental problems to assert rather than test the superiority of their preferred arrangements. Soviet claims that Marxist-Leninist societies keep their air and rivers clean (for example, Ryabchikov 1976) have a quaint ring today. Yet equally sweeping claims on behalf of other forms of social organization may be no better founded, though they have not yet been so spectacularly refuted.

To note that social position affects beliefs is not to regard all beliefs as equally valid. "That environmental fears have social correlates—even social causes—does not . . . prove them simply chimerical" (Lowenthal 1990: 130). Knowledge and intelligence are not solutions in themselves to environmental problems, but they are essential to the development of solutions. Geographers at their best subject broad generalizations and unexamined assumptions to critical scrutiny; they promote a concern for detail and a respect for fact, an appreciation of the variety of experience and a distrust of glib assertion.

So much nonsense has been written about what eminent figures from the past would think about current events—which political party would George Orwell, or Thomas Jefferson, or, for that matter, Plato vote for today, and that sort of thing—that a little more will not hurt. What would Marsh think of matters today? He would surely approve of such "honorable conquests over nature" (Marsh 1965:282) as have been won since his time: the reduction of famine (Kates and Millman 1990) and the spread of "climate-proofing" technologies that make us less vulnerable to global warming even if it should occur (Ausubel 1991). He would surely approve as well of the heed being paid, belatedly and sometimes imprecisely, to the lessons he had stated about the human relation to the earth. The most important of them is now widely accepted: that humankind is dependent upon an earth incapable of supporting infinite demands and capable both of being improved and of being damaged by the way in which it is used. That environmental catastrophes on the local and regional scales, with serious consequences for immediate human well-being, have occurred would have come as no surprise to Marsh. He could at once continue to maintain that it would be "rash and unphilosophical to attempt to set limits on the ultimate power of man over inorganic nature" (Marsh 1965:44) and repeat his warning that "the earth is fast becoming an unfit home for its noblest inhabitant, and another era of equal human crime and human improvidence, and of like duration . . . would reduce it to such a condition of impoverished productiveness, of shattered surface, of climatic excess, as to threaten the depravation, barbarism, and perhaps even extinction of the species" (Marsh 1965:43).

References

Arcury, T. A., and E. H. Christianson. 1993. Rural-urban Differences in Environmental Knowledge and Actions. *Journal of Environmental Education* 25(1):19–25.

Arcury, T. A., and T. P. Johnson. 1987. Public Environmental Knowledge: A Statewide Survey. *Journal of Environmental Education* 18(4):31–37.

Ausubel, J. 1991. Does Climate Still Matter? *Nature* 350: 649–652.

Bassin, M. 1992. Geographical Determinism in Fin-de-siècle Marxism: Georgii Plekhanov and the Environmental Basis of Russian History. *Annals of the Association of American Geographers* 82:3–22.

Blackburn, C. 1993. *New Perspectives on Environmental Education and Research.* Raleigh, N.C.: University Colloquium on Environmental Research and Education.

Boudon, R. 1982. *The Unintended Consequences of Social Action.* New York: St. Martin's.

Boyer, P. 1992. *When Time Shall Be No More: Prophecy Belief in Modern American Culture.* Cambridge, Mass.: Belknap Press.

Brookfield, H. C. 1992. "Environmental Colonialism," Tropical Deforestation, and Concerns Other than Global Warming. *Global Environmental Change* 2:93–96.

Browder, J. O. 1988. The Social Costs of Rain Forest Destruction: A Critique and Economic Analysis of the "Hamburger Debate." *Interciencia* 13:115–120.

Burton, I., R. W. Kates, and G. F. White. 1978. *The Environment as Hazard.* New York: Oxford University Press.

Butzer, K. W. 1992. The Americas Before and After 1492: An Introduction to Current Geographical Research. *Annals of the Association of American Geographers* 82:345–368.

Callendar, G. S. 1938. The Artificial Production of Carbon Dioxide and Its Influence on Temperature. *Quarterly Journal of the Royal Meteorological Society* 64:223–240.

Clark, W. C. 1986. Sustainable Development of the Biosphere: Themes for a Research Program. In *Sustainable Development of the Biosphere*, eds. W. C. Clark and R. E. Munn, 5–48. Cambridge: Cambridge University Press.

Curry-Roper, J. 1990. Contemporary Christian Eschatologies and Their Relation to Environmental Stewardship. *The Professional Geographer* 42:157–169.

Denevan, W. M. 1992. The Pristine Myth: The Landscape of the Americas in 1492. *Annals of the Association of American Geographers* 82:369–385.

Dunlap, R. E. , G. H. Gallup, Jr. , and A. M. Gallup. 1993. Of Global Concern: Results of the Health of the Planet Survey. *Environment* 35(9):6–15, 33–39.

Dunn, J. 1980. The Identity of the History of Ideas. In *Political Obligation in its Historical Context*, 13–28. Cambridge: Cambridge University Press.

Ekholm, N. 1901. On the Variations of the Climate of the Geological and Historical past and Their Causes. *Quarterly Journal of the Royal Meteorological Society* 27:1–61.

Ekirch, A. E. 1944. *The Idea of Progress in America, 1815–1860.* New York: Columbia University Press.

Elster, J. 1989. *Nuts and Bolts for the Social Sciences.* Cambridge: Cambridge University Press.

ESSC (Earth System Sciences Committee). 1988. *Earth System Science: A Closer View.* Washington, D. C.: NASA.

Glacken, C. 1956. Changing Ideas of the Habitable World. In *Man's Role in Changing the Face of the Earth*, ed. W. L. Thomas, Jr., 70–92. Chicago: University of Chicago Press.

———. 1967. *Traces on the Rhodian Shore. Nature and Culture in Western Thought from Ancient Times to the End of the Eighteenth Century.* Berkeley: University of California Press.

Gore, A. 1992. *Earth in the Balance: Ecology and the Human Spirit.* Boston: Houghton Mifflin.

Hägerstrand, T. , and U. Lohm. 1990. Sweden. In *The Earth as Transformed by Human Action: Global and Regional Changes in the Biosphere over the past*

300 Years, ed. B. L. Turner, II, W. C. Clark, R. W. Kates, J. F. Richards, J. T. Mathews, and W. B. Meyer, 605–622. Cambridge: Cambridge University Press.

Hausbeck, K. W. , L. W. Milbrath, and S. M. Enright. 1992. Environmental Knowledge, Awareness and Concern among 11th-grade Students: New York State. *Journal of Environmental Education* 24(1):27–34.

Hecht, S., and A. Cockburn. 1989. *The Fate of the Forest*. London: Verso.

Hewitt, K., ed. 1983. *Interpretations of Calamity*. Boston: Allen & Unwin.

Hirschman, A. O. 1991. *The Rhetoric of Reaction: Perversity, Futility, Jeopardy.* Cambridge, Mass. : Harvard University Press.

Hodge, C. F., and J. Dawson. 1918. *Civic Biology*. Boston: Ginn and Company.

Inglehart, R. 1990. *Culture Shift in Advanced Industrial Societies.* Princeton: Princeton University Press.

Kates, R. W., C. Hohenemser, and J. X. Kasperson, eds. 1985. *Perilous Progress: Managing the Hazards of Technology.* Boulder, Colo.: Westview Press.

Kates, R. W., and S. Millman. 1990. On Ending Hunger: The Lessons of History. In *Hunger in History*, ed. L. F. Newman, 389–407. Cambridge, Mass.: Blackwell.

Kempton, W. C. 1991. Public Understanding of Global Warming. *Society and Natural Resources* 4:331–345.

Lovejoy, A. O. 1944. Reply to Professor Spitzer. *Journal of the History of Ideas* 5:204–219.

Lowenthal, D. 1958. *George Perkins Marsh: Versatile Vermonter.* New York: Columbia University Press.

———. 1990. Awareness of Human Impacts: Changing Attitudes and Emphases. In *the Earth as Transformed by Human Action: Global and Regional Changes in the Biosphere over the past 300 Years*, ed. B. L. Turner, II, W. C. Clark, R. W. Kates, J. F. Richards, J. T. Mathews, and W. B. Meyer, 121–135. Cambridge: Cambridge University Press.

Lucas, Sir C. 1912. Man as a Geographical Agency. *Scottish Geographical Magazine* 30:449–467.

Marsh, G. P. 1965. *Man and Nature; or, Physical Geography as Modified by Human Action.* Cambridge, Mass.: Belknap Press (originally published 1864).

Merton, R. K. 1967. On the History and Systematics of Sociological Theory. In *On Theoretical Sociology: Five Essays, Old and New*, 1–37. New York: the Free Press.

Mitchell, R. C. 1989. From Conservation to Environmental Movement: The Development of the Modern Environmental Lobbies. In *Government and Environmental Politics*, ed. M. J. Lacey, 81–113. Washington, D.C.: Wilson Center Press.

Nabokov, V. V. 1964. *Eugene Onegin: A Novel in Verse by Aleksandr Pushkin, Translated from the Russian, with a Commentary*, 4 vols. New York: Pantheon Books.

Nisbet, R. 1980. *History of the Idea of Progress.* New York: Basic Books.

Parsons, J. J. 1988. The Scourge of Cows. *Whole Earth Review* Spring, 40–47.

Price, M., and M. Lewis. 1993. The Reinvention of Cultural Geography. *Annals of the Association of American Geographers* 83:1–17.

Ryabchikov, A. M. 1976. Progress of the Environment in a Global Aspect. *Geoforum* 7:107–113.

Smil, V. 1984. *The Bad Earth: Environmental Degradation in China.* Armonk, N.Y.: M. E. Sharpe.

Stern, P. C., O. R. Young, and D. Druckman, eds. 1992. *Global Environmental Change: Understanding the Human Dimensions.* Washington, D.C.: National Academy Press.

Thomas, W. L., Jr., ed. 1956. *Man's Role in Changing the Face of the Earth.* Chicago: University of Chicago Press.

Tuan, Yi-fu. 1991. A View of Geography. *Geographical Review* 81:99–107.

Turner, B. L., II, and K. W. Butzer. 1992. The Columbian Encounter and Land-use Change. *Environment* 43(8):16–20.

Turner, B. L., II, W. C. Clark, R. W. Kates, J. F. Richards, J. T. Mathews, and W. B. Meyer, eds. 1990. *The Earth as Transformed by Human Action: Global and Regional Changes in the Biosphere over the past 300 Years.* Cambridge: Cambridge University Press.

Turner, B. L., II, R. W. Kates, and W. B. Meyer. 1994. "The Earth as Transformed by Human Action" in Retrospect. *Annals of the Association of American Geographers* 84:711–715.

U.S. EPA. 1987. *Unfinished Business: A Comparative Assessment of Environmental Problems.* Washington, D.C.: U.S. EPA.

———. 1990. *Reducing Risk: Setting Priorities and Strategies for Environmental Protection.* Washington, D.C.: U.S. EPA.

WCED (World Commission on Environment and Development). 1987. *Our Common Future.* Oxford: Oxford University Press.

Weiner, D. 1988. *Models of Nature: Ecology, Conservation, and Cultural Revolution in Soviet Russia.* Bloomington: Indiana University Press.

Williams, M. 1990. Forests. In *The Earth as Transformed by Human Action: Global and Regional Changes in the Biosphere over the past 300 Years*, ed. B. L. Turner, II, W. C. Clark, R. W. Kates, J. F. Richards, J. T. Mathews, and W. B. Meyer, 179–201. Cambridge: Cambridge University Press.

PART 3

THE WORLD AS LINKED MOSAIC

7

Spatial Organization and Interdependence

Edward J. Taaffe

The Idea

The modern geographer's emphasis on spatial organization, which began with the idea of the functional region, has provided a useful perspective for viewing pressing societal questions at a number of levels. Evidence for the many ways in which spatial organization links individuals, cities, regions, and nations is around us every day. Driving through in the countryside from Milwaukee to Chicago, for instance, we notice that television antennas begin to change from north-pointing to south-pointing, and that mailbox logos shift from Milwaukee newspapers to Chicago newspapers. The network television programs we watch in either city will usually emanate from New York, Washington, or Los Angeles—and will often focus on events in far-off places such as London, Moscow, or Sarajevo. Understanding the spatial organization that links cities, regions, and nations is increasingly essential as improved transport and communications speed up the daily flows of people, goods, and information. As these linkages accelerate and strengthen, the *interdependence* of places at all geographic scales becomes increasingly clear. Decisions made in Washington reverberate in New York, in Los Angeles, in Peoria, in London, in Sarajevo—as do decisions made in Tokyo, Moscow, or Pretoria. The geographer's concern with spatial organization is reflected at all scales, within and among neighborhoods, cities, regions, and nations.

The idea of the *functional region*, first given explicit recognition in the 1920s, has had a profound effect on the ways in which geographers have approached the study of society. Initially, they focused their attention on readily visible phenomena, such as landforms, crops, and settlements, that could clearly be related to the physical environment. Regions were delimited on the basis of their internal homogeneity as regards landforms, crops, or even less visible criteria such as population diversity, income, or ethnicity. The Corn Belt, for example, might be one such region, homogeneous in the sense that every county included in the Corn Belt has some minimum acreage in corn. The idea of the functional region added *linkages* to homogeneity as a basis for identifying regions. The questions "how does this region function" or "how is this region organized" were added to the question of how the region looks or how homogeneous it is with respect to the phenomena contained within its borders.

Questions about how a region functions lead to the tracing out of the complex web of linkages within a given region and between the region and all other places both nearby and remote. The idea of the *spatial organization* of society in a place was thus added to the geographer's long-standing concern with the relation between society and the physical environment and with the distributions of phenomena within an area. This recognition of the importance of linkages and flows has led to a greater concern with the emerging *interdependence* among places—at all geographic scales. At the local scale, for example, we know that the economic and transport linkages bind city and suburban parts of a metropolitan area into a single complex system of spatial organization; yet this web of functional linkages clashes with the political cleavage of city from suburbs. Urban scholars now attribute the severity of inner-city problems to the willingness of affluent suburban residents to disassociate themselves from the economically distressed cities.

At the regional scale, the examination of changing patterns of community and intercity linkages has been tied not only to the development of central place theory, which is discussed in detail in Chapter 8. Interest in regional-scale linkages has also emerged from a growing awareness of the unintended consequences that can accompany vastly improved transportation. Better roads and widespread automobile ownership mean that people can travel much greater distances in about the same amount of time it once took to make much shorter trips. The precipitate decline and possible ultimate disappearance of the American small town is but one example of this greater mobility.

At the national level, a small set of giant metropolises dominates the spatial organization of the United States, as evidenced by corporate and banking linkages, highway traffic flows, phone calls, the growth of air hubs, the concentrations of network television, the development of computer networks and electronic superhighways, and many other indicators. Early ideas of distinct, well-delineated regions in which each major metropolitan area dominated its surroundings have been replaced by the idea of a hierarchical structure of linkages rendered more complex by the increasing strength of the largest centers at the very top of the hierarchy.

The impact of spatial organization at different scales in the Pacific Northwest is shown in Figures 7.1, 7.2, and 7.3. Hinterlands or tributary areas for lower-order activities (Figure 7.1, top) of the urban centers of the region are much smaller and more numerous than are the hinterlands for the higher-order activities (Figure 7.1, bottom). The hierarchical nature of the system is also evident in the contrast between the fine-grained local set of banking linkages (Figure 7.2, top) and the broad areas of Sunday newspaper circulation linkages (Figure 7.2, bottom). The linkages to a national system as represented by air traffic reflects still another level of complexity. The large centers, Seattle and Portland, are tied into a national system of intermetropolitan linkages (Figure 7.3). Portland is dominated by Seattle, Seattle by San Francisco, and San Francisco by Los Angeles. By the 1990s, Los Angeles was dominated by New York. Smaller Pacific Northwest centers have a mixture of air passenger linkages to regional and national centers. Although they are all dominated by Seattle or Portland, these small centers usually have more traffic to such national centers as New York, Los Angeles, or Chicago than to other small Pacific Northwest Centers.

The impact of vastly improved communications is now felt at the global level. The end of the cold war may bring about a greater rather than a lesser involvement in world political organizations, and our economic linkages have been accelerating and ramifying as major world trade agreements are completed. We have moved from a relatively simple system of trade linkages to a complex system of international and multinational corporate structures and investments, and a manufacturing structure in which foreign parts and labor are routinely involved in domestic production.

Perhaps the most fundamental effect of the functional approach to regional study and its concern with linkages has been the way linkages provide concrete evidence of our interdependence with each other no matter where we live. New York's suburbanites are not as independent of New York's central-city dwellers as they may think, and New Yorkers, in general, are not really completely independent of events in Dubuque, let alone Dallas or Los Angeles—witness the national impact

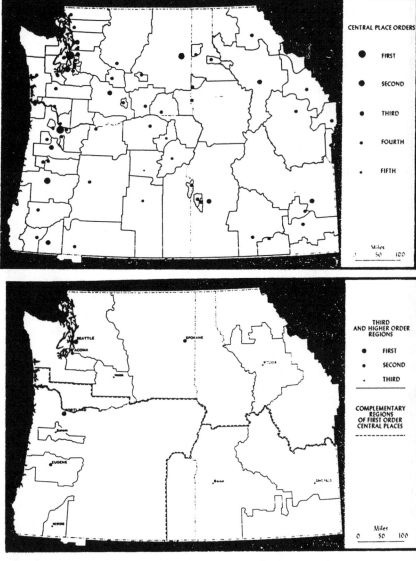

Figure 7.1. top: Hinterlands or tributary areas for lower-order activities in central places of the Pacific Northwest; bottom: hinterlands for higher-order activities in central places of the Pacific Northwest. Source: Preston (1971:148, 150).

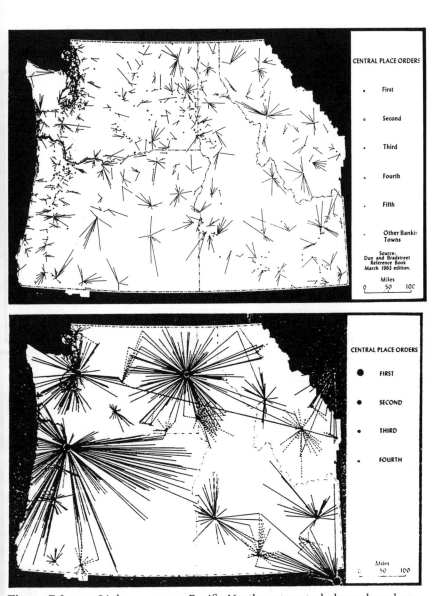

Figure 7.2. top: Linkages among Pacific Northwest central places, based on banking transactions; bottom: linkages based on Sunday newspaper circulation. Source: Preston (1971:144, 149).

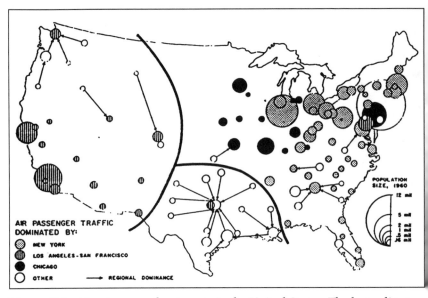

Figure 7.3. Air passenger dominance in the United States. The heavy lines divide the country into three zones of air passenger dominance. In the East, all cities are dominated by New York or by cities such as Chicago and Atlanta, which, in turn, are dominated by New York. In the West, cities are similarly dominated by Los Angeles, San Francisco, and Seattle; in the Southwest by Dallas, Houston, and New Orleans. Source: Taaffe (1970:79).

of earthquakes, floods, assassinations, and urban riots. Nor is anyone in the United States completely independent of events such as the spread of radiation from the Chernobyl explosion, the end of apartheid in South Africa, or the attempts to bring peace to the oil-rich Middle East. In the early 1990s, Americans found themselves involved to varying degrees in the events of places they had never heard of a few short years before—Kuwait, Mogadishu, Bosnia-Herzegovina.

Origins of the Functional Region

The early 1920s was a critical period in American geography. The discipline had just emerged from a protracted and divisive debate on environmental determinism and seemed to be heading in at least two general directions: "man-land" (or people-environment) relations and areal differentiation.

The environmental determinists had based much of their work on the premise that the physical environment influences human behavior.

They thought that geography's role was to look for those influences. Some went so far as to write of environmental "controls" and human "responses." By the early 1920s, the general failure of empirical evidence to support the proposed tenets had led to a general abandonment of pure environmental determinism, but an emphasis on the linkages between people and the natural environment remained. Thus, the study of man-land relations, initially articulated by Barrows (1923) as the focus of geographic study, was held to be the study of the relations between society and the natural environment without a priori notions as to either human or environmental control. This nature-society perspective remains an important one in geography. Since the 1950s, the emphasis within this tradition has been on such things as society's impact on the environment and its adjustment or adaptation to environmental hazards, ideas that are described in the previous two chapters of this book.

A second view that gained strength in the early 1920s and became the discipline's dominant definitional paradigm until well into the 1950s was areal differentiation, or the study of areas (regions). This broader, more integrative view treated a wide range of relations among people living in a given area rather than focusing only on those aspects of society that were closely linked to the physical environment. The map was the prime investigative tool because it permitted the study of diverse phenomena as they existed together in a particular place.

A key development in the history of geographic ideas during this period was a set of field conferences held in the Midwest between 1923 and 1940.[1] Fundamentally, the conferences represented an attempt by some of the country's leading research geographers to come to grips with the formidable task of deciding what were the most salient and interrelated of the overwhelming variety of phenomena to be found in any area. Many of these geographers had been trained as geologists and had a strong initial tendency to focus on the visible features of the landscape.

The land-use map epitomized the approach of the field conference geographers. Agriculture was emphasized, as were such physical features as soil, vegetation, slope, and runoff. The Montfort study (Finch 1933), which grew out of the conferences, produced a series of detailed land-use maps of a small area in southwestern Wisconsin. The analysts stressed form and often interpreted the mosaic of land uses as resulting from the geographic relationship between agricultural and physical phenomena.

Just as earlier field investigations had persuaded some geographers of the limitations of looking only at environmental relations in their effort to better understand society, so the field-conference investigations persuaded some geographers of the limitations involved in mapping just those concrete aspects of landscape that they could see and classify. The result was a move toward collecting survey or interview data and to considering each place's many different linkages to other places.

Robert Platt's (1928) study of Ellison Bay, Wisconsin, first gave explicit expression to these concerns. He went beyond the visible phenomena of the land-use map and began mapping the linkages between Ellison Bay and other places, starting with the simple delineation of a tributary area. Platt soon found, however, that linkages and orientations carried him well beyond local retail marketing ties. Many of the small community's linkages were politically based, and focused on the state capital at Madison. Others were based on wholesaling and manufacturing ties to Milwaukee and Minneapolis. Still others were reflected in rail, highway, and lake-vessel linkages to Chicago, involving recreational as well as trade ties. Migration flows of different ethnic groups resulted in still more linkages, largely from areas east of Wisconsin. The resultant maps showed a complex set of overlapping cobwebs.

Thus, by the 1930s, the area study view in geography encompassed the investigation of two types of regions. The original type visually represented by the land-use mosaic (like that used in the Montfort study) was often referred to as the *formal* region and is exemplified by a portion of a land-use map from the original Montfort study area (Figure 7.4). Here the stress is on the area's internal homogeneity, on what it looks like in terms of crops, topography, and settlement *forms*. The linkage-based type, visually represented by maps of flow lines or overlapping cobwebs, was referred to as the *functional* region. Here, the stress is on the area's organization, on how it *functions* in its internal and external ties to schools, churches, shopping centers, and major cities. It is exemplified by Brush's (1953) early study of traffic hinterlands and linkages in the same general area of southwest Wisconsin (Figure 7.4). In the same year, a more general study of the functional region was presented by Edward Ullman (1953, 1954) in which he related the idea to the broader concept of spatial interaction, which he argued was basic to all geographic work.

Since its genesis in the 1920s the idea of functional region has evolved in several directions. The basic concept of a tributary area or discrete hinterland moved to that of a hierarchy of centers and hinterlands; then to the idea of a complex network of centers, hinterland, hierarchies, linkages, and flows; and then to the broader notion of *spa-*

tial organization itself as a basis for geographic study. The types of data employed evolved from data based essentially on visual phenomena to those from a wide variety of published sources and from survey results. These data were used to develop a variety of maps based on spatial measures of orientation. In general, the linkage systems associated with the study of functional regions proved to be more susceptible to generalization than did the detailed land-use mosaics associated with formal regions. The result was a slow but steady increase in the level of abstraction used to study place.

Hinterlands, Hierarchies, and Networks

The initial emphasis was similar to that of the Montfort study, where visible, observable data were the basis for taxonomic classificatory investigations of regional "landscapes." Efforts focused on delimiting markets and other types of tributary areas and zones of metropolitan dominance. Harris's (1941) study of the role of Salt Lake City as a regional center was an excellent example. He traced out the tributary areas of a wide range of urban functions and showed how they delimited a multilayered hinterland around Salt Lake City. The term "nodal region" came into use to describe this type of focus on a single center.

It gradually became evident that only certain functions such as convenience shopping and the journey to work were susceptible to this essentially taxonomic approach. For many functions, such as those associated with durable goods or travel for recreational, religious, social, and political purposes, tributary areas could not be neatly partitioned into a honeycomb of distinct and separate hinterlands. Describing these linkages required conceptualizing more complex hierarchical systems, in which centers of varying sizes were linked in different ways for different functions. These *hierarchies* were studied by documenting people's actual trip-making for a variety of goods and services (for example, groceries, clothing, automobiles, banking) and by examining actual flows of people, goods, information, and money between places. At the national level, these studies revealed a concentration on a few very large metropolitan centers, as reflected in air traffic and banking linkages. Theoretical studies were also carried out on flows and on the ways in which city functions seemed to group into discrete, or distinctly separate, hierarchical levels.

Christaller's (1966) development of central place theory in Germany suggested an economic basis for such a hierarchical system, and the theory was applied to urban study in several different areas. One of the

EXPLANATION OF FRACTIONAL SYMBOLS
NUMERATOR

Left-hand Digit MAJOR USE TYPE	Second Digit SPECIFIC CROP OR USE TYPE		Third Digit CONDITION OF CROP
1. TILLED LAND	1. CORN (MAIZE) 2. OATS 3. HAY, (IN ROTATION) 4. PASTURE (" ") 5. BARLEY 6. WHEAT	7. PEAS (Mainly for canning). 8. SOY BEANS 9. POTATOES T. TOBACCO X. SUDAN GRASS 9/5. OATS AND BARLEY MIXED	1. GOOD 2. MEDIUM 3. POOR
2. PERMANENT GRASS LAND	1. OPEN GRASS PASTURE 2. PASTURE WITH SCATTERED TREES OR BRUSH 3. WOODED PASTURE 4. PERMANENT GRASS CUT FOR HAY		1. GOOD 2. MEDIUM 3. POOR
3. TIMBER LAND	1. PASTURED 2. NOT PASTURED		1. GOOD 2. MEDIUM 3. POOR
4. IDLE LAND	1. IS CAPABLE OF USE		

DENOMINATOR

Left-hand Digit SLOPE OF LAND	Second Digit SOIL TYPE (Wis. Soil Survey terminology).	Letter X (if indicated). CONDITION OF DRAINAGE
1. LEVEL, 0° TO 3° 2. ROLLING, 3° - 8° 3. ROUGH, 8° - 15° 4. STEEP, Over 15°	1. MARSHALL SILT LOAM 2. KNOX " " 3. " " " (STEEP PHASE). 4. LINTONIA " " 5. WABASH " " 6. ROUGH, STONY LAND	X POOR XX VERY POOR

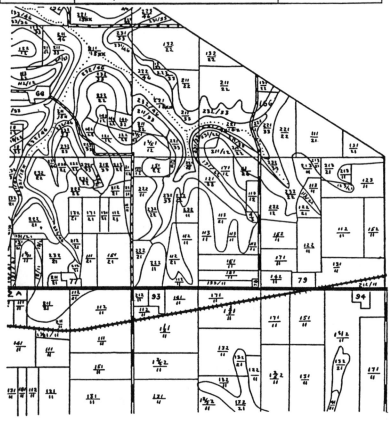

Part of the fractional complex code map of the Montfort area, another example of a mosaic of uniformity and diversity qualitatively distinguished. The Prairie Upland district in the southeastern half, the Cuesta-Escarpment district in the northwestern half. Scale: 1 inch to .4 mile.

Figure 7.4. Comparison of the two approaches to regional study, both in southwest Wisconsin. The land-use map (left) from the Montfort study represents the formal approach, stressing spatial homogeneity. Source: Platt (1959, adapted from Finch 1933). The map of linkages and hinterlands (right) represents the functional region, stressing spatial organization. Source: Brush (1953).

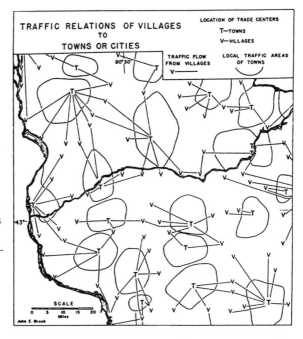

earliest explicit applications of central place theory in the United States was Brush's 1953 study of the distribution of urban settlements in southwest Wisconsin. As studies of linkages expanded beyond concerns about identifying hinterland and hierarchy to a wider examination of the varying roles of nodes of different sizes and functions in flow and linkage systems, the term network analysis became more popular. This, in turn, was related to a still more general term, spatial organization.

The spatial organization view gained strength through the 1950s and 1960s. It had always been implicit in the geographer's widespread use of the map but now had become more explicit, both empirically and theoretically. Spatial organization has remained a strong geographic tradition through the latter part of the twentieth century, taking its place along with the nature-society tradition and the area study tradition as one of the most widely used approaches by human geographers to the study of society. Most geographers, in fact, employ a mixture of these three views, varying their approach considerably with the type of problem.

In order to trace out an area's spatial organization, analysts have expanded the diversity of phenomena studied to include linkage data such as traffic flows, telephone calls, transport costs, travel times, newspaper circulation, school and retail customer records, census data on the journey to work or migration, national air passenger and banking flows, and international trade figures. Interview data on such things as trips, trip purposes, and orientations toward shopping malls or churches became common as did a greater concern with perceptions, attitudes, and preferences.

Although geographers' growing interest in the functional region was accompanied by a stress on less visible phenomena than were used to construct the land-use mosaics of the Montfort study, an even greater emphasis on mappable phenomena emerged. In addition to maps of the various field, published, and survey data, this emphasis included maps of the proximity of phenomena to cities, towns, villages, neighborhoods, and individuals in hierarchical systems and networks as well as flow maps, travel-time maps, and transaction maps. "Mental" maps were used to show how closely individuals in a given area might be linked to other areas in terms of knowledge, attitudes, or migration preferences.

Increased Level of Abstraction

The great diversity of phenomena treated in studies of the functional region and spatial organization led to more abstract generalizations about patterns of linkages and spatial organization. In addition to the abstract analyses stemming from central place theory, the ubiquitous gravity model played a role in moving geography from an initial verbal concern with functional-region linkages to an attempt at more precise expression. The volume of linkages between any two places was seen to be related to the distance between them and the size of their population. This analogy to Newton's gravitational formulation led many geographers to attempt measurements and construct ratios that, in turn, facilitated early model building and the use of statistical analysis. General structure could be expressed in relatively simple terms. This increased level of abstraction took many forms, starting perhaps with the renewed interest in central place studies in the United States, which later evolved in several directions. The early gravity formulations also evolved into a large and diverse family of spatial interaction models.

Many other types of models and other abstractions related to spatial organization and the functional region emerged after the mid-1950s.

Graph-theoretic analyses of linkages and flows made it possible to describe complex networks in relatively manageable mathematical terms. The use of mathematical programming—notably linear programming—that had been developed in operations research and econometrics, brought a normative dimension to the study of spatial organization by stressing what the most efficient pattern or spatial organization would be. Normative models of the market areas of shopping malls, for example, not only described or explained what that tributary area *was* but what it *would be* if the goal were to minimize travel or to maximize sales. The ability to define optimal tributary areas and laborsheds for retail, wholesale manufacturing, or service functions as well as for school, government, or other functions greatly increased the practical utility of the spatial organization concept. Hinterlands could be delimited so as to minimize travel, maximize sales, maximize profits, or to achieve a number of other policy goals such as greater racial balance in schools. Planners could also use these models to suggest optimal flow patterns and transport mode mixes within given areas. Such models enabled policy makers to experiment with the probable consequences of a number of various policy alternatives before actually committing funds to them.

The spatial organization idea soon came to include the functional region as represented by linkages and the formal region as represented by the mosaic. As the external linkages of the region change, the economic and social patterns within the region change accordingly. The economic idea of regional specialization, for example, is based on the relation between the two. Figure 7.5 illustrates how a change in linkages can bring about changes in land use. When there are no linkages (Figure 7.5A) the land around City X has the same wide range of agricultural activities as does that around City Y—even though the land around City X may be the best in the world for wheat and the land around City Y may be the best in the world for sugar beets.

Once a single linkage is established (Figure 7.5B), regional specialization begins and the land-use patterns change. Because the wheat can be produced more cheaply around X than around Y, farmers around Y stop raising wheat and Y meets its needs by importing wheat from X. Farmers around X increase their production of wheat and stop raising sugar beets. Specialization increases as both cities expand their linkages to provide access to a world port (Figure 7.5C).

The recent growth of geographic information systems or GIS (see Chapter 3 in this volume) has facilitated the interweaving of linkage-based and mosaic-based perspectives in the study of spatial organiza-

Figure 7.5. The diagram illustrates how a change in linkages brings about a change in land use. In (A) the agricultural regions around each isolated city are similarly diversified and are focused on the subsistence needs of the cities. In (B) the cities are linked by effective transportation and begin to specialize: X in corn, Y in sugar beets. In (C) the transport linkages are expanded to include national and world markets. Specialization has intensified and cities X and Y have become interdependent not only with each other but with many other cities. Source: Taaffe and Gauthier (1973:35).

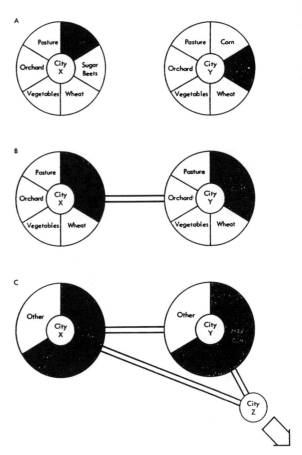

tion. GIS involves the computerization of huge data bases so as to facilitate the comparison of linkage and flow information with information about the characteristics of areas. Such comparisons have become possible at levels far beyond those previously available. GIS also permits the incorporation of abstract models of various types of linkage systems into map studies that include both land-use mosaics and actual linkage systems.

The Idea's Impact

The idea of studying ties and linkages between places, which started with the simple identification of functional regions and broadened to a concern with spatial organization in general has resonated well beyond the confines of geography. We are all becoming increasingly

aware that distance is not a simple concept but a complex idea underlying an expanding variety of spatial organizations. We are also becoming aware of the equally complex web of interdependencies that
accompanies each technologically driven change in distance relations.
The entire system of linkages and interdependencies differs markedly
with scale. What is true of an individual or community may not be at
all true of a region or nation.

The idea of spatial organization has helped us to realize that distance as it affects spatial organization is not merely a question of geometric miles between places, but a considerably more complicated
idea that is often quite subjective in nature. Distance may be expressed
in terms of time, cost, ratios, perceptions, or orientations, and it has
different meanings with different technologies. Canals, highways, jet
aircraft, FAXes, and e-mail all produce different linkages and spatial
organizations. New technologies do not free us from the constraints of
distance; we merely become subject to different and more complex
distance relationships. Consider the following examples. Using the ex-
pressway to travel between parts of a large city often requires less
travel time than does a direct route involving more traffic lights.
Routes may be chosen to maximize amenities rather than to minimize
driving time. A large city is often better connected to other large but
distant cities than it is to smaller nearby cities. Farm villages become
absorbed into elaborate hierarchical systems.

As we examine our linkages and orientations toward other places,
other complications emerge. The increased importance of the highest
levels of the intermetropolitan hierarchy becomes clear. In entertainment and the arts, a bipolar New York-Los Angeles structure appears
to be increasingly prominent; in national politics, we are more likely to
be familiar with our U.S. senator than our state senator or district
representatives. One might argue that distance has been replaced by
frequency of media mention as a criterion for proximity. The development of the "information superhighway" calls for using still other criteria in examining linkages and flows. Instead of thinking that the
effects of distance have been annihilated by rapid advances in transport and communications technology, we now realize that these effects have just become more complex and interesting.

Above all, the greatest significance of the spatial organization and
functional region idea lies in its explicit recognition of interdependence. The changing structure of linkages among places reflects and
reshapes the interdependencies of these places. The changes take place
simultaneously at local, regional, national, and global levels, and, in
the process, raise a number of difficult policy questions.

At the local or metropolitan level, as noted earlier, the arbitrary distinction between suburb and city (simply a political boundary) obscures the interdependent functional linkages of the metropolitan area as a whole. This political division leads to a surprisingly large number of conflicts and inequities. As Kozol (1991) notes, the property-tax basis for funding education results in suburban schools being far better funded than the central-city schools in the same metropolitan area, despite the clearly greater needs of the latter. Supreme Court rulings on the "separateness" of city and suburb led to the exemption of most suburban schools from bussing programs. This, in turn, accelerated "white flight," which, in turn, acted to intensify already strong city-suburb social and economic contrasts. The artificial political barrier between functionally integrated city-suburb components feeds a cycle that further exacerbates fiscal disparities. As education and crime-control services in the central city are perceived as inadequate, individuals and businesses move out to suburban jurisdictions, where these services are perceived as adequate. This reduces the tax base of the central city, leaving it with the alternatives of raising tax rates or reducing services. Either action further intensifies the flight from the city, which cycles back and further reduces the city's tax base, which, in turn, further weakens services and thereby initiates still another cycle of flight.

At the regional level, changing linkage structures threaten the very existence of small towns and cities as unchecked market forces intensify concentrations in large metropolitan complexes. In addition, discrepancies between the spatial organization of U.S. industry and that of the fifty states pose serious problems to certain regions. Much U.S. industry is organized on a national basis and, with the increased mobility of improved transportation and communication, it has become relatively footloose as regards location. Industry has been attracted to states with low wages, restrictive union legislation, and a willingness to offer tax and other incentives to these industries even if it should mean continued state underfunding of such things as welfare, Medicaid, and education. This mobility has created disadvantages for industrialized states in the Northeast and Midwest as many manufacturers have moved south and west. Despite the fact that many industries are organized nationally, federal legislation regarding relevant aspects of tax and labor policy has been difficult to enact because of its control at the state level.

At the national level, we have yet to evaluate the consequences of the hierarchical system currently being intensified by airline deregulation and a hub-and-spoke network that leads to greater economic concentration in a small group of just eight or nine super cities. To complicate matters further, evidence suggests a steadily increasing concentration on the upper end of the hierarchy: The nation's network of air passenger flows has become increasingly concentrated on New York during the past half century. In 1940, five dominant centers could be identified; in 1989 there were only two—New York and Dallas. In 1940, approximately one-half of the largest air-passenger-generating cities in the United States had more traffic to New York (or a city dominated by New York) than to any other city. By 1990, this had increased to nearly 90 percent.

The relation between the functional region and the formal region becomes particularly significant at the global level. In addition to the trade relations discussed earlier, the necessity of considering the relation between local cultural mosaics and the webs of interdependence at all scales is particularly urgent today. If we concentrate only on local cultural mosaics of ethnicity, religion, or nationality, we forget linkages to the rest of humanity through economic ties, political organization, information flows, and so forth. Conversely, if we concentrate solely on external linkages, we risk stressing standardization and homogeneity while failing to consider the legitimate needs and aspirations of locally diverse populations.

For example, the post-Cold War conflicts in former Yugoslavia are primarily based on local ethnic rivalries. Some possible solutions involve linkages to regional organizations such as the North Atlantic Treaty Organization (NATO) as well as global linkages to the United States and the United Nations. To be effective, however, any solution must consider the relation of these external linkages to the local relations among Serbs, Croats, and Bosnians as well as among Roman Catholics, Eastern Orthodox Catholics, and Muslims.

Thus the idea of spatial organization has had the effect of broadening our concern with societal problems. The geographic implications of Donne's "no man is an island" have been carried from a physical analogy to an abstraction with clear empirical expression. In the study of interdependencies, geographers need not confine themselves to relations to the physical environment. Geographic perspectives as reflected in spatial organization can be quite useful even in cases where no aspect of the physical environment is involved. But spatial organization should not stand alone. Networks based on economic, social, and political interrelations are intertwined with the logical interdependencies described in Chapters 4 and 6.

Thus the idea of looking beyond the immediate local landscape of a few small areas in Wisconsin that began well over a half-century ago continues to unfold and lead us in many diverse and constantly changing directions. We have only begun to appreciate the complexity of flows, linkages, and orientations that mark the spatial organization of our cities, our regions, our nation, and our world.

Note

1. A small group of geographers gathered together every summer for a number of years to exchange ideas about new approaches to geographic study. These approaches were then applied to the field investigation of a relatively few locations in the Midwest. For a description of these conferences, see James and Mather (1977).

References

Barrows, Harlan H. 1923. Geography as Human Ecology. *Annals of the Association of American Geographers* 13:1–14.

Brush, John E. 1953. The Hierarchy of Central Places in Wisconsin. *Geographical Review* 43:380–402.

Christaller, Walter, 1966. *Central Places in Southern Germany*, translated by C. W. Baskin. Englewood Cliffs: Prentice-Hall.

Finch, Vernor, C. 1933. Montfort—A Study in Landscape Types in Southwestern Wisconsin. *Geographic Society of Chicago Bulletin* 9:15–40.

Harris, Chauncy D. 1941. *Salt Lake City: A Regional Capital.* Ph.D. dissertation, University of Chicago, Chicago.

James, Preston E., and E. C. Mather. 1977. The Role of Periodic Field Conferences in the Development of Geographical Ideas in the United States. *The Geographical Review* 67:446–462.

Kozol, Jonathan. 1991. *Savage Inequalities: Children in American Schools.* New York: Crown Publishing.

Platt, Robert S. 1928. A Detail of Regional Geography: Ellison Bay, Community as an Industrial Organism. *Annals of the Association of American Geographers.* 18:8E1–126.

———. 1959. Field Study in American Geography. *Department of Geography Research Paper* 20(61):105–114. Chicago: University of Chicago Press.

Preston, Richard E. 1971 The Structure of Central Place Systems. *Economic Geography* (47):136–155.

Taaffe, Edward J. ed. 1970. *Geography.* Englewood Cliffs: Prentice-Hall.

Taaffe, Edward J., and Howard L. Gauthier. 1973. *Geography of Transportation.* Englewood Cliffs: Prentice-Hall.

Ullman, Edward L. 1953. Human Geography and Area Research. Abstract, *Annals of the Association of American Geographers* 43:238–239.

———. 1954. Geography as Spatial Interaction. *Annals of the Association of American Geographers* 43:54–60.

8

Nested Hexagons: Central Place Theory

Elizabeth K. Burns

As Labor Day approaches, we travel with our children to a regional mall to buy back-to-school clothing, but for a pair of socks a small neighborhood store will usually suffice. We do our weekly grocery shopping at a neighborhood supermarket or a discount warehouse, but when all we need is a half-gallon of milk or a bag of Doritos, we go to the nearest convenience store or gas station food mart. For a routine medical checkup we go to a nearby doctor's office or clinic, but for major surgery we travel to a large regional hospital.

Similar patterns can be found in the largest metropolitan areas and smallest rural towns. The staples of daily life—gas, snacks, bank transactions, mail boxes—are available in large numbers and are widely distributed. Convenience stores, gas stations, local bank branches, money machines, and mail boxes are familiar elements of the urban landscape. Central libraries, fire and police stations, regional malls, and medical complexes of hospitals and offices are equally familiar but are fewer in number, farther apart, and located at sites that are widely accessible to people living throughout a large region. These facilities serve many people, but no one person needs to have them nearby on a daily basis. Such services are located in large building complexes, often on large sites, located at major street intersections or close to freeway interchanges so as to be accessible to many people.

Each firm or company locates its stores and offices to reach the widest possible audience of customers and clients. In urban settings, people can usually travel easily to other, competing stores offering the same products. Clusters of fast food outlets, gasoline stations, and chain grocery stores occur on busy streets as each establishment competes for a share of the passing clientele. Ideally, the market area of each McDonald's restaurant, for example, does not overlap the market area of another McDonald's outlet. Companies like Wal-Mart and fast-food franchises locate their outlets in places where enough customers are regularly available and local competition may be weak (Graff and Ashton 1994).

What Is Central Place Theory?

Central place theory provides a comprehensive approach to understanding the spatial organization of human settlements, specifically the location of consumer goods and services. Its wide influence lies in recognizing predictable relationships among consumers, firms, and urban places. These relationships are visible in different geographic landscapes, in different cultural settings, and in different historical periods. They are evident in Ottumwa, Iowa, as well as in New York City. In addition, central place relationships provide the basis for specific private decisions and public actions.

Central place theory emerged as an essential component of North American urban economic geography in the 1960s, when metropolitan areas and the urban service economy in the United States were growing rapidly. Central place theory was born when a few individuals identified and organized some key principles describing efficient location and travel patterns. These theoretical insights were quickly seen to have direct practical value, first, in explaining rural settlement patterns and, later, in understanding and shaping land-use patterns in urban areas, where these concepts have been widely used in retail location and marketing, medical geography, and public facility location.

How do the decisions of businesses and of individual consumers fit together in actual store and office locations? Central place theory offers a connected system of concepts that recognizes the tight linkages between consumer travel and points of sale. Table 8.1 summarizes these concepts.

The *market area* of any service provider is the area in which its consumers are located, whether the provider is a delivery store selling pizza or a school providing education. The geographic size of a market area depends on things like population density and consumer purchasing power; market areas will be smaller, for example, in densely settled or in high-income areas than in areas of low density or low income. The regular activities of customers comprise a *threshold* of purchasing

Table 8.1. Central Place Concepts

Central place	A town or city that provides a common location for obtaining goods and services.
Central place hierarchy	The grouping of central places by population size and types of goods and services.
Market area	The hinterland or surrounding area served by a central place.
Range	The distance a customer will travel for a specific good or service.
Threshold	The purchasing power of customers and consumers that supports provision of a good or service.

power or customer support that varies with local population density, disposable incomes, and culture. A pizza delivery outlet is profitable only if it serves an area having enough people with the purchasing power and desire to buy pizzas. When these threshold conditions are met at a particular location, a pizza store becomes viable there. Similarly, a school's service area requires enough families with children to provide at least the minimum school size to satisfy the local school board.

From the provider's perspective, the *range* describes how far a customer will travel to acquire a particular good or service. People are usually willing to travel only short distances for goods and services they purchase frequently—gas, snacks, or cash from an automatic teller. Less frequently needed purchases or services merit greater effort and a longer trip—a medical specialist, a car purchase, or a mortgage loan. A business operator or city planner will try to identify business and service locations where a large number of customers or users can meet on a regular basis. Ideally, each business or service location is reached by individual trips that require a low average travel time when all customers and users are considered. Clearly, the greater the number of service locations (such as branch libraries) in a particular area, the shorter the distance people will have to travel on average to reach one.

Central places are towns and cities that provide a common location for obtaining goods and services. Customers find multiple stores and services in central places, while the firms located there reach many more customers than they would if they were not located with other outlets. A drive through the countryside or a look at a map with place

Figure 8.1. The 1961 central place hierarchy of southwest Iowa. Source: Berry (1967); reprinted by permission of Prentice-Hall.

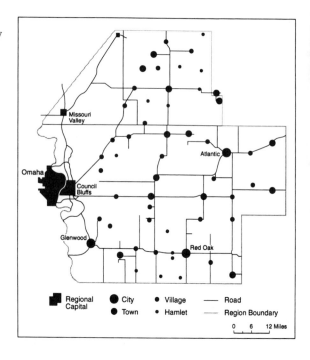

names confirms these observations. Small towns serve not only the town residents but also a dispersed regional population. Small towns are more closely spaced in areas of concentrated farming than in areas of ranching or mining. A few large cities serve as regional centers; their sites reflect advantages of convenient road, railroad, and air access serving their surrounding region and its small towns. The largest metropolitan areas—New York, Los Angeles, and Chicago—serve national and global markets.

This *hierarchy* of central places describes a pattern in the number and regular spacing of towns and cities. The pattern reflects the evolution of a balance between, at one extreme, providing goods and services in every small town and, at the other extreme, requiring that customers or suppliers travel to and from a single national metropolitan area. At a regional scale, large cities are widely spaced, few in number, and serve large areas within which smaller cities and towns serve more local needs. A pattern like this existed in southwest Iowa in 1961. Figure 8.1 describes this five-level central place hierarchy as it applies to the twin cities of Omaha, Nebraska, and Council Bluffs, Iowa, located on the Missouri River, which serve as the regional center for smaller cities, towns, villages, and hamlets (Berry 1961).

Together, these concepts provide an integrated explanation of the location, size, economic characteristics, and spacing of market centers in functional regions. Central place theory's lasting appeal is this coherent interpretation of universal location principles based on observation of actual consumer behavior, provider decisions, and settlement locations. The theory's widespread impact occurs, however, through application to urban and regional issues.

The planning of new communities is often grounded in central place principles. When large undeveloped parcels in urban areas, such as the former Irvine Ranch in Orange County, California, are developed, neighborhoods and commercial services can be planned and located without the constraints of an earlier street plan or existing land parcels and buildings. Regional shopping, services, and office centers serve as central business districts for these communities and are designed to minimize consumer travel and to provide efficient market locations for stores, medical services, and offices. Today, suburban freeway intersections provide these accessible locations for new communities.

The new town of Columbia, Maryland, built in the 1960s halfway between Baltimore, Maryland, and Washington, D.C., is a prominent example. The single town center is the community focus, providing the equivalent of a small town's main street. The town center is marked by its large buildings, many commercial outlets, and easy access on the local street network from each neighborhood (Figure 8.2). Each neighborhood also has a central commercial center, but the neighborhood centers are not specialized and customers have little need to travel directly between them for daily goods and services. Served by curvy local streets with only a few connections to major arterials connecting the neighborhood to the town center and peripheral highway, each neighborhood is self–contained. This street design reinforces the primacy of the single town center, which offers stores and offices not available in the neighborhoods. Office and industrial sites attracting workers from Columbia and surrounding towns are located at the town's periphery away from residential neighborhoods.

Urban and regional planning applications illustrate the wide acceptance of central place concepts and provide some cautions on their use. Planning a community on ideal principles is difficult, especially because consumer preferences and business conditions change more quickly than the built environment. In Columbia, for example, daily goods and services are often bought outside the residents' neighborhoods on the way to and from work. The neighborhood purchasing power that planners envisioned is not fully available, therefore, to support the neighborhood centers.

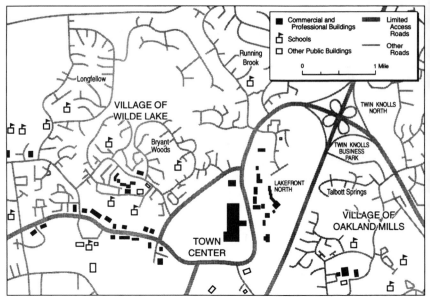

Figure 8.2. The location of the town center and neighborhood centers in Columbia, Maryland. Source: © The Rouse Company (n.d.)

In the 1980s U.S. metropolitan areas adopted central place principles in targeting the largest employment and service centers as the basis for regional land use and transportation planning. Los Angeles, among other cities, chose a centers–based approach to give a focus to land-use planning for its many districts and neighborhoods (Hamilton 1986). The City of Phoenix, Arizona, adopted a similar approach when nine urban villages were designated for preferential public and private development support in the 1986 General Plan (Burns 1988; Fink 1993) (Figure 8.3). The number and location of urban village centers were influenced as much by the location of existing shopping malls and developable vacant regional center sites as by any analysis of local market areas and travel patterns.

Regional settlement systems illustrate prescriptive planning actions with a longer history. Small rural settlements with transportation links to a larger service community were designed in the newly developed Lachish region of Israel and Dutch polderlands reclaimed from the North Sea since World War II (Constandse 1963; Takes and Venstra 1960). Recent Dutch planning policy for these polders has shifted toward supporting fewer, large settlements and toward maintaining less agricultural land in larger parcels. These trends are a response to in-

Figure 8.3. The City of Phoenix, Arizona, Urban Village Plan. Source: City of Phoenix Planning Department (1991:6).

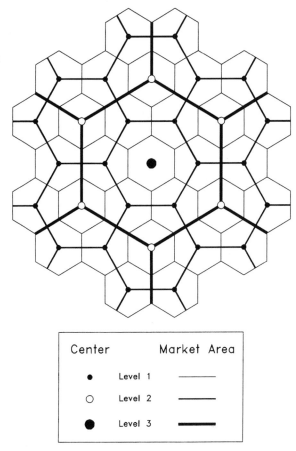

creased agricultural mechanization, increased commuting to urban jobs, and widespread automobile use on improved highways.

Origins

Central place concepts originated in the regional settlement analyses of two German geographers and were adapted by North American geographers for urban settings. These theoretical concepts, which explained the spatial structure of service activities and were supported by geometric illustrations of ideal economic landscapes, were a revelation for geographers ready to expand the descriptive approaches that prevailed at the time.

During the late 1950s and early 1960s, urban and economic geographers shifted their focus from rural societies and landscapes to urban societies and the rapid growth of American cities after World War II. Like other social sciences, urban and economic geography underwent a quantitative revolution, in which quantitative methods and abstract models were prized. The focus of geographic research and understanding shifted away from regional and landscape studies toward the enunciation of general principles, with less emphasis on the unique character of each place.

By the 1950s, functional area studies, described in Chapter 7, had defined key concepts that would be integrated into central place theory. The broad region served by and dependent upon a city was well understood to be its hinterland or tributary area. The national system of cities was a hierarchical system analogous to a wedding cake with several tiers. Its base consisted of many small hamlets and villages supporting larger cities and regional capitals, which in turn supported the few, largest metropolitan areas. Activities such as migration, air and rail travel, industrial production, and retail sales were all connected within this network. Brian J. L. Berry and William L. Garrison (1958) identified a hierarchy in the number and location of commercial activities in towns in central Washington state. Berry (1962) conducted studies of shopping trips by rural and urban residents in southwest Iowa, the farming region and central place system illustrated in Figure 8.1.

Central place theory emerged in the seminal work of a German geographer, Walter Christaller, whose book, *The Central Places of Southern Germany*, was originally published in 1933 but not translated into English until the mid–1960s (Christaller 1966). His ideas, and those of his follower, August Lösch (1954), did not have a wide impact on thinking in the United States until the 1960s, although Ullman (1941) ably summarized Christaller's ideas in English as early as 1941. Initially, Christaller defined and combined the concepts of the threshold of purchasing power, the consumer's range, the central place role of urban settlements, and the nested relationship of smaller places within the market areas of larger places. Lösch recognized that Christaller's model assumed that each central place provided every good and service and encouraged monopolies for service providers in each market area. Lösch's ideal model addressed these concerns and resulted in a complicated pattern of regional settlements in the market area of each major central place. Together, Christaller's and Lösch's analyses focused on the geometric and mathematical characteristics of sets of urban places that together make up a central place system.

The value of Christaller's original contribution cannot be overstated.

He built his analysis on knowledge of a set of specific places, the regional settlements of Southern Germany. His unique insight, however, was to lay out a set of principles that could apply to all places. The mental shock of his worldview on some geographers was substantial. Suddenly, the familiar topic of rural settlement patterns was seen from a new angle. Christaller's insight enabled geographers to see urban places in the context of integrated locational analysis.

How did Christaller make this conceptual leap? He took the powerful step of outlining the common concepts that create a consistent, coherent central place structure. He then showed how these concepts would interact in an ideal landscape, one that had uniform population density, terrain, soil, and mineral resources and where a transportation network served all areas equally. The lack of detail about the particular characteristics of specific regions is an advantage in describing uniform travel and economic decisions. Each consumer is expected to travel to the nearest place where a good or service is provided. In addition, an entrepreneur is expected to provide that service whenever enough customers exist to support it profitably.

Christaller's approach results in an ideal economic landscape with a clear geometric pattern of nested hexagonal market areas (Figure 8.4). Businesses locate in smallest centers to serve dispersed rural settlers; these hamlets are spread apart and reflect minimum customer travel distances to reach enough customers to support their businesses. A regular pattern emerges. One of Christaller's ideal geometries describes nested hexagonal market areas shown in Figure 8.4. The largest center (level 3) serves a market area equal to the market area served by three of the next smallest (level 2) centers. Each of these centers in turn serves a market area equal to the market area served by three of the smallest (level 1) centers.

This nested honeycomb pattern shows how market areas are combined by having each larger central place serve a market area three times larger than the next smaller market area. In this geometric pattern each central place's market area does not overlap with adjacent market areas and no outlying customers are left unserved. Christaller assumed that these small market areas nest within larger market areas. Larger central places support specialized goods and services that require more customers to generate a profit. Multiple outlets exist for the same type of services in these larger towns.

When an effort is made to measure the significance of an idea, one approach that works well in practice is the degree of controversy that surrounds the question of who first stated it. There is considerable testimony to the importance of central place theory, therefore, from

Figure 8.4. Christaller's ideal central place hierarchy showing nested hexagons. Each central place serves the market area of three smaller central places. Source: Berry and Parr (1988:57); reprinted by permission of Prentice-Hall.

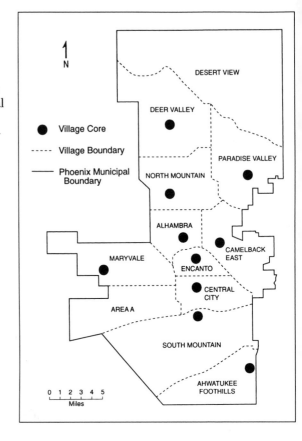

the number of writers credited with anticipating Christaller's work in whole or in part. These authors are now seen as precursors to Christaller in the same way that Leif Erickson preceded Columbus: premodern Arab scholars such as al-Khorizmi (Gould 1985:14) and al-Muquaddasi (Berry, Conkling, and Ray 1976:226–227); Richard Cantillon (Dawson 1969; Fairbairn and Barr 1974) in eighteenth-century France; a host of nineteenth-century writers including A. H. Muller von Netterdorf (Dawson 1969), J. Reynaud (Robic 1982; Lepetit 1988), Leon Lalanne (Getis 1962), M. Goubert de Ger and Elisee Reclus (Dunbar 1978), the British Census Commissioners of 1851 (Freeman 1961), the American S. H. Goodin (Abbott 1981: 115; Hamer 1990: 132) and, in the early twentieth century, Grant Allen and H. G. Wells (Blouet 1977), J. Tischler (Kellerman 1979), and Charles Josiah Galpin (Ullman 1941). In some instances, to be sure, only a partial insight or shared interest in the spatial pattern of settlements existed. In other cases, the resemblance is substantial. Yet it was assuredly Christaller's formulation of a complete system of central places that caught the attention of geographers and planners and influenced their thinking. It is not inaccurate or unjust to describe him as the author of the theory.

By the time that Brian J. L. Berry's 1963 book *Commercial Structure and Commercial Blight* appeared, central place concepts were widely known in geography (Berry 1963, 1967). His critical insight was identifying systems of commercial centers within urban areas as well as at the regional scale. He linked central place concepts to the actual business patterns in American cities, using Chicago as his main example. Berry's shift in focus to a single urban area, with its variety of commercial centers and districts, ensured that central place theory would not be applied only to rural and regional analysis, but would become fundamental to our understanding of modern urban spatial organization.

Later extensions of central place theory provide more realistic descriptions of economic conditions. Business providers make locational decisions on the basis of incomplete information, for example, and consumers of different household types and cultural backgrounds vary in their travel behavior and service preferences. Central place theory remains influential today because it provides a coherent explanation for an essential component of all urban population and economic growth: the service economy.

The theoretical power of the original central place concepts was clearly shown by their prescriptive systems of service activities. A constant tension therefore exists between the high level of generality that clarifies these enduring principles and the need to be aware of the specific conditions of particular places. Geographers do not find hexagonal market areas in the landscape, but the overall principles do shape the spatial organization of economic behavior.

Central place concepts contribute directly to current regional settlement issues. The Great Plains region developed with dispersed agriculture on individual farms; by the early twentieth century a network of small and large central places had emerged there. John Hudson (1985) suggests that a kind of "folk location theory," roughly equivalent to Christaller's guided the town site planners of the post–Civil War American railroads. Small towns and their surrounding populations have been declining, however, as conditions have changed in the areas served by local central places. Wheat farming has become mechanized, land holdings have been consolidated, and less labor is now needed to produce the crops; customer travel times have shrunk as automobiles replaced horse-drawn wagons for shopping trips. This changing population base and transportation technology has removed nearby consumers that used to support the smallest towns. Berry and Parr (1988:30) note that, "as densities decline, centers of a given level serve fewer people" and services concentrate at higher levels in the central place hierarchy. The decline of population and central places is so widespread that a Buffalo Commons, on the scale of a multi-state national park, has been proposed to allow return of buffalo herds to depopulated portions of the Great Plains (Popper and Popper 1987).

Similar issues of declining services and unmet local population needs emerge in urban settings as well, especially in low-income neighborhoods. Central place concepts show how urban commercial locations reflect variations in nearby consumer purchasing power and income. South Central Los Angeles is one example of a low-income central city community whose daily services are not fully met. According to the 1990 census, its population of about one million people had a median income of $20,820, 32 percent below the county median of $30,525. The population is 49.2 percent black and 44.8 percent Latino. Although there is one store for every 203 residents in Los Angeles County, there is only one store for every 415 residents in South Central (*Los Angeles Times* 1991). Moreover, there is one service business— movie theater, auto repair shop, law firm, or hotel—for every 290 residents, while the county ratio is one service business for every 103 residents.

These comparisons provide a starting point to identify imbalances across a metropolitan area in a community's aggregate income, purchasing power, and the actual goods and services provided. South Central Los Angeles has many check-cashing outlets but few full-service banks, many fast-food outlets but few restaurants with varied menus. This community faces private and corporate business resistance to investing in low-income black and Latino neighborhoods. The lack of basic goods and services, however, has a strong negative impact on the people who live and work there. The few stores offering wide ranges of goods and services can charge higher prices because distant competitors can be reached only by long car or bus trips.

Evolution and Expansion

Within geography, central place theory has undergone some important changes in character and emphasis. It was principally used at first to describe and explain existing patterns of city and town location. In time it became apparent that, in many cases, it could not account for existing settlement patterns. Cities developed for many reasons (for example, industrialization) not considered within central place reasoning (Page and Walker 1991:285–286). Today it is used mainly in reference to patterns of service and market locations, rather than city or settlement locations. Central place theory tends to be used less as a key to describing existing patterns than as a normative or prescriptive basis for identifying what improvements can and should be made in those patterns. If it does not show the world as it is, it shows the world as it ought to be. These shifts in the application of central place theory have distanced it from current research in urban geography, where it once flourished, but have made it an important tool in the domain of planning, in both the industrialized and the less developed world, and its power to change the world has increased accordingly.

Applications of central place theory in the developing world have flourished since the 1970s (Rondinelli and Ruddle 1978; Rushton 1988). Planners using Christallerian concepts have recognized the importance of a dense hierarchical network of markets and service centers, especially in rural areas, to meet the needs of a dispersed population. In many settings such a network is poorly developed; there are too few lower-order centers, in particular, to meet people's needs, and the hierarchy of service centers is often top heavy, with the largest centers being too big. The perspectives offered by central place theory have helped geographers and planners to recognize the disadvantages for a country of what the geographer Mark Jefferson (1939) called a "primate city," that is, a dominant national urban center much larger than any other city and inordinately significant in national life. When Jefferson recognized this form of urbanization, his intent was to praise it as giving nations a center and a focus. Today, in large part because of central place theory's insights into the usefulness of a balanced and dispersed settlement and service hierarchy, urban primacy is generally seen as a problem that contributes to severe regional inequalities. The active promotion of secondary centers is seen as an important measure for dealing with these inequalities.

Contemporary applications of central place concepts in North American cities recognize existing conditions of varying consumer knowledge, current travel conditions, and imperfect business decisions. Central place concepts have wide influence in the areas of retailing and marketing and in the location of public facilities and medical services. These topics share common locational questions. Where should stores, offices, and hospitals be located to serve a dispersed population?

Individual store owners and firms use the concepts of central place threshold and range to locate retail outlets. Every entrepreneur wants a monopoly in the market area that supports his or her facility (Collins

1989; Guy 1991); entrepreneurs will try to avoid competition from other stores by selecting sites at a distance from competitors or by concentrating many outlets at popular locations to limit a competitor's access to the full customer market. The many choices available to urban shoppers mean that no one store or mall can expect to capture all the purchasing power generated by its nearby population. More than thirty years ago, David L. Huff (1963) confirmed that customers living closest to Park Forest Plaza, a Chicago suburban shopping center, were the most likely to shop there. The percentage of residents who shopped at this mall decreased as distance from the mall increased.

Today retailers are increasingly likely to specialize to compete with new shopping locations—suburban discount warehouses, urban outlet stores, upscale malls in historic districts—that draw customers from wide market areas. Widespread automobile use contributes to a decline in the smallest stores within easy walking distance; neighborhood centers that are accessible by car gain from serving this larger customer base (Handy 1993). In metropolitan Seattle, the retail hierarchy includes two central business districts (downtown Seattle and suburban Bellevue) and suburban centers that specialize at a metropolitan scale by providing furnishings (Southcenter) and antiques (Greenwood) (Morrill 1988).

A national retailing example illustrates the rapid evolution of central place marketing applications. Innovative marketing approaches that eliminate the need for local travel still must identify the purchasing power of specific customer groups. Mail catalogs, the telephone, and television directly link customers and retailers without requiring trips to local stores. Companies are well aware of different national and local consumer markets for their products, however, and market accordingly. Using data from U.S. Census documents, sales records, and customer surveys, private marketing firms analyze retail geography at the scale of individual postal codes (zipcodes). These firms classify potential customers by lifestyle preferences for products such as cars, food, magazines, and entertainment. The Claritas Corporation identifies forty national neighborhood types (Weiss 1988), vividly described in catchy phrases ranging from Blue Blood Estates (the wealthiest suburbs, including Beverly Hills-California's zipcode 90210) to Public Assistance (the impoverished central city neighborhoods and rural counties, including Watts-California's zipcode 90002).

Central place theory also undergirds the provision of public services to consumers in urban and rural service areas. Providing public services from a central facility raises questions of optimal travel time and routing within a service area. These questions are faced by emergency service dispatchers from 911 calls, police departments, and fire stations, all of which need to identify an appropriate response, locate the vehicle and staff to respond, and dispatch the unit to a specific address where there is an emergency. Here, mapping programs and geographic information systems provide up-to-date maps showing street networks, specific addresses, and the location of emergency units and facilities for immediate response as well as planning purposes.

When consumers must travel to a central location such as a school, the site should be chosen to reflect the spatial distribution of users and their travel routes and modes. Every community has a familiar hierarchy of many elementary schools, the service areas of which are nested within middle and high school service districts. Classic neighborhood design locates elementary schools in the center of residential subdivisions where traffic is light and children can easily reach the school on foot or by bicycle. Ideally, the subdivision provides the number of children to support an elementary school.

This close relationship between the adjacent residential area and a primary school breaks down for larger schools that must draw from larger areas or magnet schools with specialized courses of study. It also breaks down when busing is required to achieve racial balance within schools and when the family composition of a neighborhood changes. A school may be closed, for example, if the nearby neighborhood no longer has enough elementary school children. An unneeded elementary school can sometimes become a senior citizen center or condominium complex, but the new use is not always suited to the original location.

In a similar way, the pattern of medical facilities in every metropolitan area illustrates the presence of a distinct hierarchy, from the level of an individual doctor, to neighborhood clinics, and finally to full-service hospitals with every possible kind of equipment and specialization. Major hospitals serve as the focus for doctors' offices, medical equipment suppliers, nursing homes, and outpatient clinics. These clusters often emerge in an unplanned fashion in central city locations near older hospitals, where existing buildings and services are a major investment. In the United States, new medical campuses in suburban locations are designed for a mix of medically related activities (Urban Land Institute 1989).

This hierarchical medical model is not permanent, however. No single medical facility operates in isolation from other nearby facilities and the patients who use its services. Maintaining multipurpose facilities with inflexible locations incurs high costs when the characteristics of nearby patient populations shift. The loss of middle-income residents from central city neighborhoods has meant that the older hospitals located there now have a lower-income population base nearby. Suburban population growth supports decentralized branch hospitals. Competition occurs from other medical facilities, including outpatient clinics providing innovative surgical or testing technologies. These concerns led to an evaluation of facilities on a regional basis in Great Britain, where suburban areas and new towns were underfunded in hospital and community-level care (Mohan 1988).

Influence

Any idea that has been so often used has inevitably often been misused. The history of central place theory's employment for dubious ends begins with the theory's author himself; Christaller spent the early 1940s in a Berlin office trying to apply his concepts to the replanning of settlement patterns in Nazi-occupied Poland (Rossler 1989). Central place concepts have often been applied, especially in Third World settings, in an overly mechanical and simplistic way, relying on a rearrangement of spatial patterns to resolve problems that have much deeper-seated social origins (Bromley 1983; Gore 1984). Its use in rural development in Africa and Latin America has in some cases produced poor results or even been counterproductive, sharpening rather than alleviating existing inequalities (Southall 1988). But to exclude from our thinking any idea capably of being misapplied or applied badly would be to give up thinking entirely, and to abandon central place theory for some doubtful episodes in its past would be to lose an exceptionally stimulating and insightful set of perspectives. Although they have sometimes helped change the world for the worse, they can readily help to change it for the better.

Central place concepts remain influential because they explain important activities and behavior in human settlement patterns, namely the provision of goods and services. The key insight is that the complexity of human landscapes is not random, but is partially ordered by the spatial interaction of consumers and service providers in large and small places. Central place principles remain influential also because they are intellectually resilient. Although societal conditions change—for example, personal mobility increases and settlement densities shift—these concepts remain powerful explanations for the way in which goods and services are provided in central places.

Central place concepts have influenced related fields such as marketing and urban planning that incorporate these locational principles into their theory and practice. As we have seen, the ideas in central place theory allow predictions for future actions in areas where these principles are best understood—retail and commercial sites, medical services, school and public facility location.

Central place theory increases our awareness of efficient and inefficient spatial patterns and behavior. Clearly, consumer and firm location decisions can lead to spatial monopolies, underserved communities, and duplication of facilities. An imbalance of service providers can exist, such as too many or too few doctors in one community. The activities best explained by central place theory are a large component of all urban economies. Central place concepts are constantly being applied and extended in ways that touch multiple aspects of our daily lives. Each service innovation and consumer preference shift reinforces the value of a continuing awareness of central place concepts.

Note

Thanks to Bill Meyer for his suggestions on several portions of this chapter.

References

Abbott, C. 1981. *Boosters and Businessmen: Popular Economic Thought and Urban Growth in the Antebellum Midwest.* Westport, Ct.: Greenwood Press.

Berry, B.J.L. 1962. *Comparative Studies of Central-place Systems.* Washington, D.C.: Office of Naval Research, Geography Branch.

———. 1963. *Commercial Structure and Commercial Blight.* Chicago: University of Chicago.

———. 1967. *Geography of Market Centers and Retail Distribution.* Englewood Cliffs, N.J.: Prentice-Hall.

Berry, B.J.L., E. C. Conkling, and D. M. Ray. 1976. *The Geography of Economic Systems.* Englewood Cliffs, N.J.: Prentice-Hall.

Berry, B.J.L., and W. L. Garrison. 1958. The Functional Bases of the Central Place Hierarchy. *Economic Geography* 34:145–154.

Berry, B.J.L., and J. B. Parr. 1988. *Market Centers and Retail Location: Theory and Applications.* Englewood Cliffs, N.J.: Prentice-Hall.

Blouet, B. W. 1977. H. G. Wells and the Origin of Some Geographic Concepts. *Area* 9:49–52.

Bromley, R. 1983. The Urban Road to Rural Development: Reflections on US-AID's 'Urban Functions' Approach. *Environment and Planning* A 15:429–432.

Burns, E. 1988. Urban Planning Within the Salt River Valley. In *Metro Arizona,* ed. Charles Sargent, 165–166. Scottsdale, Ariz.: Biffington Books.

Christaller, W. 1966. *Central Places in Southern Germany,* translated by C. W. Baskin. Englewood Cliffs, N.J.: Prentice-Hall. Originally published in 1933.

City of Phoenix Planning Department. 1991. *General Revised Plan for Phoenix.* Phoenix: City Planning Department.

Collins, A. 1989. Store Location Planning: Its Role in Marketing Strategy. *Environment and Planning* A 21:625–628.

Constandse, A. K. 1963. Reclamation and Colonization of New Areas. *Tijdschrift voor economische en sociale geografie* 54:41–45.

Dawson, J. E. 1969. Some Early Theories of Settlement Location and Size. *Journal of the Town Planning Institute* 55:444–448.

Dunbar, G. S. 1978. *Elisee Reclus: Historian of Nature*. Hamden, Ct.: Archon Press.

Fairbairn, K. J., and B. M. Barr. 1974. Acknowledging the Past: Richard Cantillon's Pattern of Urban Settlement Location. *Area* 6:208–210.

Fink, M. 1993. Toward a Sunbelt Urban Design Manifesto. *Journal of the American Planning Association* 59:320–333.

Freeman, T. W. 1962. *One Hundred Years of Geography*. Chicago: Aldine.

Getis, A. 1962. A Report on the Work of Leon Lalanne. *The Professional Geographer* 14(3):27.

Gore, C. 1984. *Regions in Question: Space, Development Theory and Regional Policy*. London: Methuen.

Gould, P. 1985. *The Geographer at Work*. London: Routledge & Kegan Paul.

Graff, T. O., and D. Ashton. 1994. Spatial Diffusion of Wal-Mart: Contagious and Reverse Hierarchical Elements. *The Professional Geographer* 46:19–29.

Guy, C. M. 1991. Spatial Interaction Modelling in Retail Planning Practice: The Need for Robust Statistical Methods. *Environment and Planning* B 18:191–203.

Hamer, D. 1990. *New Towns in the New World: Images and Perceptions of the Nineteenth-century Urban Frontier*. New York: Columbia University Press.

Hamilton, C. 1986. What Can We Learn from Los Angeles? *Journal of the American Planning Association* 52:500–507.

Handy, S. 1993. A Cycle of Dependence: Automobiles, Accessibility, and the Evolution of the Transportation and Retail Hierarchies. *Berkeley Planning Journal* 8:21–43.

Hudson, J. C. 1985. *Plains Country Towns*. Minneapolis: University of Minnesota Press.

Huff, D. 1963. A Probabilistic Analysis of Shopping Center Trade Areas. *Land Economics* 39:81–90.

Jefferson, M. 1939. The Law of the Primate City. *Geographical Review* 29:220–226.

Kellerman, A. 1979. Location Theory Anticipated. *Area* 11: 347–348.

Lepetit, B. 1988. *Les villes dans la France moderne (1740–1840)*. Paris: Albin Michel.

Los Angeles Times. 1991. Retail Exodus Shortchanges Consumers. *Los Angeles Times*, A1 and A38–A39, November 24.

Lösch, A. 1954. *The Economics of Location*, translated by W. J. Woglom and W. F. Stolper. New Haven: Yale University Press. Originally published 1939, reprinted by Wiley, New York, 1967.

Mohan, J. 1988. Restructuring, Privatization and the Geography of Health

Care Provision in England, 1983–1987. *Transactions of the Institute of British Geographers*, new series 13:449–465.

Morrill, R. L. 1988. The Structure of Shopping in a Metropolis. *Urban Geography* 8:97–128.

Page, B., and R. Walker. 1991. From Settlement to Fordism: The Agro-industrial Revolution in the American Midwest. *Economic Geography* 67:281–315.

Popper, D. E., and F. J. Popper. 1987. The Great Plains: From Dust to Dust. *Planning* 53(12):12–18.

Robic, M. -C. 1982. Cent ans avant Christaller, une theorie des lieux centraux. *Espace Geographique* 11:5–12.

Rondinelli, D., and K. Ruddle. 1978. *Urbanization and Regional Development: A Spatial Policy for Equitable Growth.* New York: Praeger.

Rossler, M. 1989. Applied Geography and Area Research in Nazi Society: Central Place Theory and Planning, 1933 to 1945. *Environment and Planning* D 7:419–431.

Rushton, G. 1988. Location Theory, Location-allocation Models, and Service Development Planning in the Third World. *Economic Geography* 64: 97–120.

Southall, A. 1988. Small Urban Centers in Rural Development. *African Studies Review* 31(3):1–15.

Takes, C.A.P. and A. J. Venstra. 1960. Zuyder Zee Reclamation Scheme. *Tijdschrift voor economische en sociale geografie* 51:162–167.

Ullman, E. L. 1941. A Theory of Location for Cities. *American Journal of Sociology* 46:853–864.

Urban Land Institute. 1989. *Development Trends 1989,* special issue. Washington, D.C.: Urban Land Institute.

Weiss, M. 1988. *The Clustering of America.* New York: Harper & Row.

9

Megalopolis:
The Future Is Now

Patricia Gober

Nestled along the western shores of Lake Michigan between Milwaukee and Chicago is Kenosha, Wisconsin. Thirty years ago, Kenosha was a bustling, industrial town of some 65,000 persons. The Charlie Crows and Al Fedemeiers of Kenosha worked in the town's automobile factories and metal fabricating plants. The community was freestanding in the sense that its residents worked, shopped, and played largely within its own boundaries. Today, Kenosha has been enveloped by the great midwestern megalopolis stretching from Pittsburgh to Milwaukee. Its economic base now reflects its position at the outer fringes of this vast network of urbanized areas rather than its historic local industrial capacity. The sons and daughters of Charlie Crow and Al Fedemeier now work at the Dairyland Greyhound Park (a dog racing facility), at the I-94 Tangier outlet mall (the community's major tourist attraction), and in construction and other activities serving the steady influx of Chicago commuters. In contrast to its relatively self-contained past, Kenosha and its residents are now intimately linked into a large urban network of people and places. Megalopolis has arrived in Kenosha.

The idea of megalopolis was popularized by the French geographer, Jean Gottmann, around 1960. The term can be used in two ways: as the proper name for Gottmann's original study area—the urbanized Northeast of the United States—and as a generic term for the coalescence of metropolitan areas into a continuous network of urban development. Megalopolis symbolized an enlarged scale of urban life, new forms of spatial organization, changing modes of economic behavior, and the advent of information as the raw material of urban economic life. Gottmann went so far as to argue that megalopolis signaled a turning point in the history of human settlement.

This chapter explores how and why did megalopolis become such an important idea, how it was used to explain the momentous geographic, economic, political, and social changes in American cities, what effects it had on the discipline of geography, how it entered the popular lexicon, and what it says about the future of urbanization worldwide.

From "Megalopolis" to "megalopolis"

People had long been aware that some kind of urban corridor was emerging along the northeastern coast of the United States, encompassing Boston, New York, Philadelphia, Baltimore, and Washington. Journalists called it "Boswash." During the 1950s, while associated with the Institute for Advanced Studies at Princeton University, Jean Gottmann made a thorough study of this corridor and wrote the definitive book, *Megalopolis: The Urbanized Northeastern Seaboard of the United States.* He concluded that it was more than just a chain of contiguous cities that would probably fuse over time. What Gottmann saw that others had not was a new form of urban organization that might serve as a prototype for future urban development. He named it Megalopolis, Greek for "great city" to mark its status as a new kind of urban structure, but one with deep historical roots (Gottmann 1961).

Megalopolis represented a change in the scale of urban life from the single metropolis to a network of metropolitan areas growing together and connected by movements of people, goods, and, to an ever larger degree, information. Boston, New York, Philadelphia, Baltimore, Washington, D.C., and their suburbs were no longer self-contained city systems but subsets of a much larger urban realm that stretched 600 miles from north to south (Figure 9.1). The single metropolis with a dominant city center, its suburbs, and surrounding countryside was giving way to a more polycentric and nebulous organization. Gottmann observed that some cities were losing their former centrality and were becoming suburbs or satellites of communities that had neither the size nor the functions traditionally associated with central places. Downtown activities were moving out of the city's core into the periphery. Rural and urban land uses were merging, making urban life look more rural and rural life look more urban.

But Megalopolis was more than just an increase in the size and scale of urban life and more than a change in the form of cities. It repre-

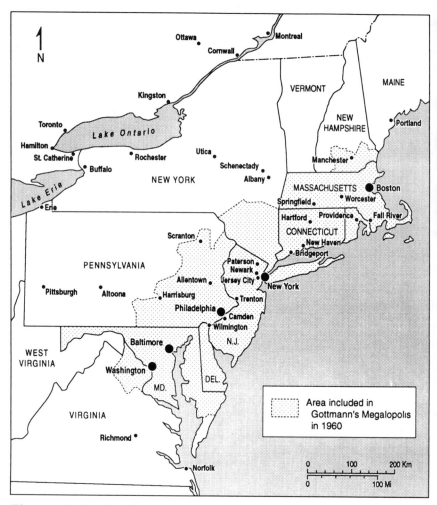

Figure 9.1. Megalopolis as defined by Jean Gottmann in 1961. Source: adapted from Swatridge (1971:4).

sented an increase in the intensity of the interactions that are the basis of urban life. Economic specialization and a refined division of labor require efficient communication and transportation. Gottmann noted the extraordinary density of activities and movements of all kinds including highway traffic, airline travel, human migration, newspaper circulation, and telephone calls (Figure 9.2). That Gottmann characterized Megalopolis's residents living in different states and hundreds

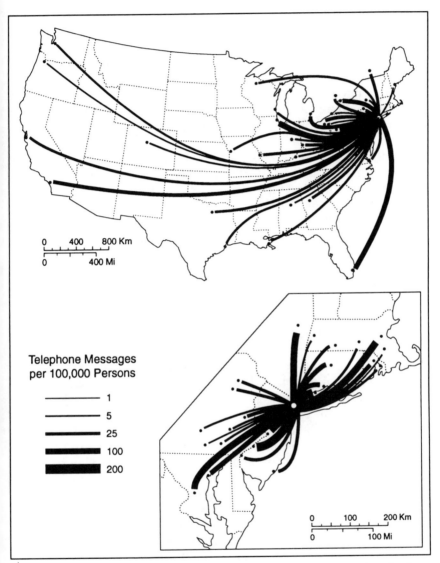

Figure 9.2. Average number of daily telephone messages per one hundred thousand persons between New York and selected cities in 1958. Source: adapted from Gottmann (1961:590).

of miles apart as "close neighbors" speaks to the importance of these functional interactions (as opposed to geographic proximity) in defining closeness.

A third feature of Megalopolis was its role as an "economic hinge" linking its national hinterland with the outside world. Cities historically attempted to increase their influence by bringing more territory into their regional or national spheres. New York, for example, markedly improved its trading position during the early nineteenth century with the construction of the Erie Canal. Before the Erie Canal was opened in 1825, connections from the interior to the east coast were extremely costly and entailed a cumbersome water route down the Ohio River to the Mississippi, then to New Orleans, and finally on sailing ships to eastern seaboard ports. High costs on this route made long-distance shipment of bulk products unprofitable. Farmers in the Ohio Country would distill their grains into whiskey to reduce the weight of their product and increase its value enough to pay the expensive shipping fees.

Before 1820, east coast cities like New York, Boston, Baltimore, and Philadelphia fiercely competed to determine which one could incorporate the largest segment of the interior into its market area, thereby enhancing its status as an economic hinge. Although Baltimore was first to breach the Appalachian Barrier with the National Road in 1818, New York was first to develop a cost-efficient linkage via the Erie Canal. The canal established an uninterrupted water connection between New York City and the Great Lakes via the Hudson River and 364 miles of engineered waterway stretching from Troy, New York, on the Hudson to Buffalo on Lake Erie (Figure 9.3). Transportation costs on the Erie Canal were one cent per ton-mile compared to 13 cents on the National Road (Bigham and Roberts 1962). The drastic reduction in long-distance shipping costs afforded by the Erie Canal allowed New York to tap as its hinterland the midwestern territory around the Great Lakes and, after the railroads were completed, parts of the Great Plains as well. By 1840, New York's population had mushroomed, and the city had achieved dominance over the nation's economic landscape (Pred 1966; Taaffe and Gauthier 1973) (Figure 9.4).

Gottmann's economic hinge notion was an extension of the traditional hinterland notions of urban economic growth. Megalopolis ("Boswash") flourished as a result of the size and wealth of its national hinterland but also from its ability to link this hinterland with the rest of the world. Megalopolis derived power and dominance from its position as a center of international finance, as headquarters for multinational corporations, and as a major exporter of advanced services.

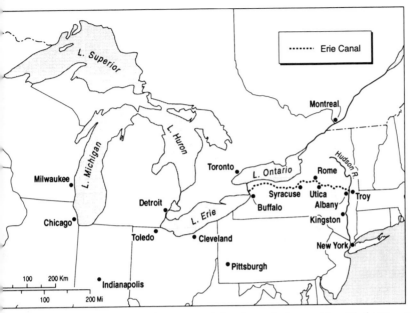

Figure 9.3. The Erie Canal connected the Great Lakes to New York City via the Hudson River.

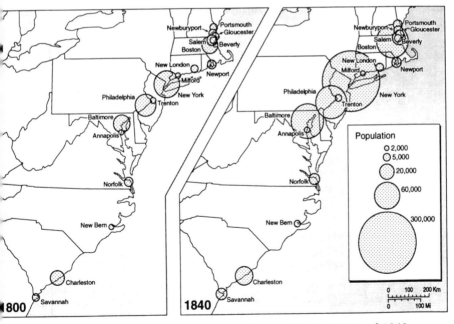

Figure 9.4. Population of major eastern seaboard cities in 1800 and 1840. Source: adapted from Pred (1966:188).

With the emergence of a global economy, the value of these external relationships is patently obvious to us today. The idea, however, was far less apparent and more innovative in the late 1950s when Gottmann first described Megalopolis.

Commensurate with its flourishing role as an international entrepôt, Megalopolis represented a significant shift in urban employment patterns. Fewer people were engaged in the production and handling of goods, and more in services. The shift in employment from factories to offices and from blue- to white-collar occupations was dubbed the "White Collar Revolution." This revolution spawned a new family of economic enterprises called "quaternary activities." These highly specialized services involve information analysis, research, and decision making and include administration, finance, insurance, public relations, marketing, education, government, scientific research, and the performing arts.

In keeping with the presence of quaternary activities, Megalopolis became a "transactional space"—a place where ideas are exchanged. Emphasis was on the production and distribution of information rather than of goods. Conventional wisdom at the time was that improvements in telecommunications would lead to the dispersal of settlement. Modern telecommunications would allow people to fulfill their professional and personal needs without having to live in dense, compact cities. From their dispersed residences, individuals would be able to communicate with the rest of the city and, indeed, with the rest of the world. With the collapse of time and space, the need for physical proximity, the very basis for urban life, would diminish.

Gottmann (1983) argued that better telecommunications technology would not necessarily mean that cities will become obsolete. It depends upon how people decide to use that technology. Television has made it possible to view almost all forms of entertainment at home, and yet people still choose to attend concerts, sports events, the theater, and the cinema. The Home Shopping Network and mail-order catalogs have not yet undermined the downtown and the mall as places to shop and as public gathering places. Telephone conferencing and videophones have not eliminated the need for business people to meet face-to-face. Gottmann contended that isolation and dispersal are not the inevitable outcomes of a transactional society. Transactional activities and telecommunications technology actually promote more communication and more face-to-face contact, which, in turn, lead to enhanced agglomeration and bigger cities. Activities not dependent upon personal contact, such as routine manufacturing and wholesaling, are pulled toward the periphery and are replaced by transactional activities in the center.

Gottmann's meticulous description of Megalopolis, emphasizing the forces underlying modern urban development, begged the question: Were there other megalopolises? Were there other places of similar size, with dispersed and polycentric configurations, intense internal ties, crucial linkages to the global economy, and specializations in advanced services? Most certainly there were, and geographers and others began to see them all over the world (Isomura 1969; Hall 1971; Doxiadis 1974; Gottmann 1976; Leman and Leman 1976).

The closest parallel to Megalopolis was the Tokaido region of Japan, a conurbation stretching more than five hundred miles along the Pacific Coastline from Tokyo to Osaka. The word megalopolis, transliterated into a Japanese phonetic script called katakana, became megaroporisu. In 1967 two geographers, Ishimizu Teruo and Kiuchi Shinzo, published a shortened translation of Gottmann's *Megalopolis*. Two years later, Isomura Eiichi, an urban sociologist, produced a detailed description of the Tokaido megaroporisu. Architect Tange Kenzo maintained that the Tokaido Megaroporisu, like its American counterpart, was not just a chain of autonomous metropolitan regions but a single urban realm joined by "mutual interconnectedness" (Hanes 1993). The most visible symbol of this interconnectedness was the Shinkansen, or "bullet train." Completed for the 1964 Tokyo Olympics, the bullet train moved passengers between Tokyo and Osaka at astonishing speeds; travel time between the two cities fell from eight to three hours. The bullet train epitomized Tokaido's status as a megalopolis and as one of the premier urban regions of the world.

In 1976, Gottmann christened another four megalopolises and spotted three potential candidates. The former included the Great Lakes region from Quebec to Milwaukee, a north-south urban corridor in England, northwestern Europe from Amsterdam to the Ruhr, and the urban constellation centered on Shanghai. The three urban areas that appeared to be coalescing fast enough to qualify for megalopolis status were the Rio de Janeiro–São Paulo complex in Brazil, southern Europe centered on Milan, Turin, and Genoa, and the coast of California from San Francisco to San Diego (Gottmann 1976). Fifteen years after Gottmann described the original Megalopolis, the term had come to represent a generic process and had become a regional moniker.

The Intellectual Context

The idea of megalopolis bucked a powerful antisuburban bias among urban theorists of the time and challenged a number of long-standing traditions in geography. Gottmann's *Megalopolis* was one of four seminal books about cities and American society published in 1961. The others were Lewis Mumford's (1961) *City in History*, Jane Jacobs's (1961) *The Death and Life of Great American Cities*, and Daniel Boorstin's (1961) *The Image*. One of megalopolis's main features was its ability to suck vast expanses of rural territory into its sphere of influence, creating, in the process, acre upon acre of suburban sprawl. Boorstin, Mumford, and Jacobs criticized suburban life as shallow, derivative, and artificial. They argued that urban functions require a level of population density and a diversity of activities that are diluted by the low-density, homogeneous suburban development featured in megalopolis. To Boorstin, Mumford, and Jacobs, the suburban sprawl associated with megalopolis was inimical to the very essence of city life: the high densities needed to support a stimulating and creative community.

Among these writers Gottmann stood alone in emphasizing the positive aspects of megalopolis. The size and complexity of a megalopolis, he observed, enlarged the need for people to communicate and created more opportunities for social exchange. Moveover, suburban sprawl relieved the suffocating crowding of industrial cities. Gottmann's vision of megalopolis was far more in touch with the aspirations and behaviors of common people, people who did not share Mumford's and Jacobs's enchantment with dense, chaotic, urban concentrations. After World War II, young GIs and their families flocked to Levittown and similar suburban communities where newly built single-family homes were available and affordable. Far from signaling the doom of urban civilization, megalopolis was, in Gottmann's view, to be celebrated as a crowning symbol of technological and economic achievement. Gottmann's colorful portrait of megalopolis as a "stupendous monument erected by titanic efforts" (Gottmann 1961:23) and as "an attraction for successful or adventurous people from all over America and beyond" (Gottmann 1961:15) put a distinctly positive spin on a process that had been derided by the outspoken and pessimistic urbanists of the time.

Within American geography, *Megalopolis* marked a growing concern with urban issues. The discipline was slow to shift away from its traditional focus on rural landscapes and regional inventories. In retrospect, this reluctance seems curious in light of the fact that the nation's population was urbanizing at a breakneck pace. The percentage of the U.S. population living in urban areas had risen from 33 percent in 1900 to 70 percent in 1960 on the eve of the publication of *Megalopolis* (U.S. Bureau of the Census 1975). Since then the percentage has continued to increase, albeit more slowly, reaching 75 percent in 1990 (U.S. Bureau of the Census 1992).

In the midst of this sweeping transformation of the nation's economic landscape and settlement system, the three dominant paradigms in American geography at mid-century either ignored urban studies altogether or constricted them to an extremely narrow set of research questions. Two of these traditions focused explicitly on the relationship between people and the natural environment. In the first, Carl Sauer and the so-called Berkeley School of cultural geography sought to understand human use and modification of the landscape, with a strong bias toward rural areas and premodern cultures. One of Sauer's interests was the history and diffusion of agriculture (locating the centers of plant and animal domestication and the paths by which new technologies are diffused) (Solot 1986). His penchant for forging links with anthropology and his emphasis on historical, as opposed to contemporary processes, sparked little interest in cities or urban affairs (Marston et al. 1989).

A second research theme at mid-century grew out of a long-standing concern for human-environment interactions at the University of Chicago. Described in Chapter 4 of this volume, this theme stressed the idea of human adjustment—the way people adjusted to living in hazardous areas. Early writings focused on how and why people occupied floodplains and how they adjusted to flooding. This work served as the springboard for studying other natural and technological hazards and helped to shape U.S. public policy on managing floodplains and other hazardous areas. Although the scale of development in megalopolis makes it especially vulnerable to natural hazards, this line of research was not explicitly urban in its orientation. Cities were of interest as potentially hazardous sites not as geographic entities to be studied in their own right.

Regional geography, the third powerful research tradition of the 1950s, stressed the synthetic nature of geography and produced a series of regional inventories that drew together aspects of the climate, topography, biogeography, agriculture, industries, and population of an area. The settlement system—the location of cities, their size, growth, function, and relationship to hinterland—was a piece of this regional inventory, but cities were rarely the focus of study in regional geography.

It is probably not surprising that the inspiration for megalopolis came from someone outside the parochial mindset of American geography, from someone taking a fresh look at the geography of North America and being struck by the seemingly boundless forces of urbanization. Although *Megalopolis* was, in some ways, a regional inventory of the northeastern United States, it was much more. Gottmann focused on the processes of urban growth, on the forces that created and sustained this unprecedented concentration of people, economic activity, political power, and cultural influence. His explicit urban focus fired young geographers' interest in cities and, along with other landmark ideas of the 1950s and 1960s, laid the groundwork for urban geography as a burgeoning and distinct subfield. By 1984 the Urban Geography Specialty Group had become the largest specialty group in the Association of American Geographers.

The timing of *Megalopolis*'s debut also coincided with a changing emphasis in geography from regional synthesis and analysis to spatial organization. During the 1960s and early 1970s, geography, like the other social sciences, was undergoing a "quantitative-scientific revolution" that entailed shifting away from regional inventories and a concern with the unique aspects of places toward a focus on spatial analysis and the features that places have in common. Megalopolis was a transitional idea in this disciplinary switch. Its intellectual roots lay in the detailed regional description of the urban Northeast, but at the heart of the idea was a strong spatial sense. Gottmann was concerned with spatial patterns of land use, urban density, transportation, and communications; with the spatial reorganization of megalopolis's commercial and manufacturing structure; and with the inherently spatial processes of concentration and deconcentration. Megalopolis provided the foundation from which more universal laws of urbanization were built.

Megalopolis had strong ties to emerging ideas about functional regions in economic geography, which Edward Taaffe describes in Chapter 7 of this volume. The connectedness and interdependence among places within megalopolis, expressed in the density of commuting, the intensity of telephone communications, intercity movements of passengers and goods, and the whole notion of a transactional city, were arguably its most distinctive and innovative features. That cities were physically growing together was obvious even to the lay observer. That they were growing together in terms of the movements of people, goods, and information was a more abstract and complex idea.

Why Was Megalopolis so Influential?

Megalopolis was a riveting idea because it helped to explain what was really happening in the world at the time. It provided a framework for geographers to explain changes in population distribution and the settlement system. It offered related disciplines a vision of human settlement that needed to be planned for and regulated. And most significantly, megalopolis could easily be translated into the everyday lives of people who lived in megalopolis. Many people understood, in a subliminal way, that the scale of their life patterns had enlarged and that the center of gravity of their cities had shifted, but no one had articulated the "big picture" until *Megalopolis* was published. The book and its ideas were widely reviewed in the popular press, and megalopolis soon entered the language and thoughts of American society.

Although the popular and journalistic notion of megalopolis fixated on cities growing together and on the expanded scale of urban life, the key features of a megalopolitan economy were more influential in an academic sense. *Megalopolis* drew attention to fundamental changes in the urbanization process including the shift from manufacturing to services, the emergence of quaternary activity, information as the basis for urban growth, and the changing orientation from regional to global economic systems. Recognition of these changes led to a growing concern in economic and urban geography with the study of producer services, corporate headquarters, office location, and transactional cities.

The concept of megalopolis also influenced related academic fields. Within planning, megalopolis stimulated a change in the scale of analysis from the city to more regional-level planning. Economist Lauchlin Currie (1976) argued that the basic design of our great cities was still the small town clustered around a central plaza, village green, or square—a design that was increasingly outmoded as cities grew in size. Recognizing the inevitability and desirability of increasing urbanization, Currie sought to define an urban form that would simultaneously preserve the economies of bigness, reduce the need for vast amounts of urban space being devoted to transportation, preserve open space near dense concentrations of people, and decrease inequalities between rich and poor. His solution was "cities-within-cities," in which a large metropolitan area is made up of compact, walkable, planned communities of sufficient size to be true cities. Each of these communities within the larger urban system was expected to hold between 400,000 and 500,000 residents, larger than the typical British New Town or American planned community.

Nowhere was the influence of megalopolis on planning greater than in Japan. Following Gottmann's optimistic lead, the Japanese tended to idealize megaroporisu as a symbol of Japan's economic and technological success. In 1965 architect Tange Kenzo developed a blueprint for the future of Japan in his book, *Image of the Future of the Japanese Archipelago*. Noting the extraordinary postwar population growth and concentration of political and economic power in the Tokaido Belt, he chided those who clung to the anachronistic notion of free-standing metropolitan regions. Meeting the transportation and communications needs of people living at a megalopolitan scale of life became the imperative of Japanese planning. Infrastructure took precedence over social and environmental goals in this imperative (Hanes 1993).

Gottmann's attention to the green spaces in Megalopolis and his articulation of the challenges inherent in managing these spaces caught the interest of landscape architects. Gottmann noted that a wooded appearance predominated in Megalopolis in the 1950s. Almost 50 percent of the region was forested. This figure represented a substantial increase over the region's pre-Megalopolis forest cover. Ironically, urban sprawl was associated with reforestation. Formerly cultivated land was being abandoned and was reverting to its natural wooded state at a faster rate than new land was being converted to urban uses. The challenge for landscape architects was to identify how best to structure the region's woodlands and to identify what forms of nature would best satisfy the physiological and psychological needs of Megalopolis's residents. In a keynote address to a 1968 symposium of the New York Botanical Garden, Gottmann stressed the enhanced importance of nature in Megalopolis, where residents spend most of their time at densities of several thousand persons per square mile in residential areas and more than one hundred thousand per square mile in the areas where they work. Megalopolis "requires us to think of cities as an integral part of a wider system within which the urban is not to be opposed to the rural, but interwoven in a completely new fashion" (Gottmann 1970:65).

Gottmann urged Americans to let go of the "urban" and the "rural" as a fundamental dichotomy (Gottmann 1961:343). The notion of urban land uses as homes, offices, factories, stores, and streets in contrast to rural images of forests, farms, and wetlands was increasingly outmoded as the two types of land uses intermingled in megalopolis. Suburban housing developments, along with stores and offices, replaced farmlands just outside the core of megalopolis; the growth of second homes in the outer periphery reinforced the integration of urban and rural land uses as well as urban and rural lifestyles. The reemerging woodlands interspersed among the region's homes, offices, and factories further undermined traditional notions of what was distinctly urban or rural.

Megalopolis also was a pivotal building block for the field of Ekistics, the science of human settlement. During the early 1970s the Greek planner Konstantinos Doxiadis described the tendency for the world's population to consolidate into ever larger units, from towns to metropolises, megalopolis, eperopolis or continent-size agglomerations, and eventually to ecumenopolis, a worldwide system of linked settlements. In keeping with his futuristic perspective, Doxiadis developed a scenario for the physical layout and topographical structure of ecumenopolis in the year 2100, taking into account the presence of established human settlements, the physical geography of world regions, historical and cultural traditions, and present trends and patterns of urbanization. The European ecumenopolis was to be focused on the London-Paris-Amsterdam axis with tentacles stretching out to include the continent's major urbanized areas (Figure 9.5). By no coincidence, this core "looks toward" the eastern seaboard of the United States, clearly anticipating the strong linkages that will emerge between these two great megalopolises. Doxiadis and his followers maintained that the distribution of the world's population in the early 1970s was in transition from the small towns of the past to Ecumenopolis, the city of the future. Within this framework, megalopolis was a model for understanding past changes in the nature and scale of urban life and for predicting future trends (Doxiadis 1974).

Megalopolis Today

Just as urbanists came to expect ever larger levels of population concentration, bigger metropolitan areas, and more megalopolises, forces of centralization began to dissipate around 1970. Between 1970 and 1980 nonmetropolitan areas in the United States grew faster than metropolitan areas, and among metropolitan areas, growth was inversely related to size. The largest metropolitan agglomerations were experiencing the slowest rates of population growth and, in some cases, were even declining (Champion 1989b). First recognized in a much heralded paper by demographer Calvin Beale (1975), these trends were then given context and force by geographer Brian Berry (1976), who coined the term "counterurbanization" to describe them. According to Berry, counterurbanization stemmed from the shift from industrial to postindustrial modes of production and from popular preferences for small towns, lower densities, and environmental amenities. Berry's counterurbanization thesis resonated in the United States and abroad,

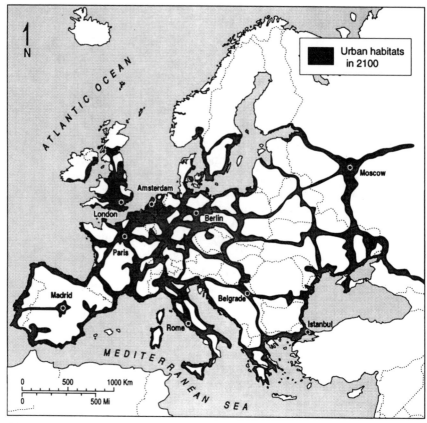

Figure 9.5. Zone of urban habitation in Europe as foreseen by Doxiadis in 1974. Source: adapted from Doxiadis (1974:371).

where social scientists found similar trends in many of the industrial and industrializing nations (Champion 1989a).

Much to the surprise of counterurbanization proponents and to the dismay of those trying to predict future trends, the 1980s brought a stunning reversal in counterurbanization tendencies in the United States, Great Britain, and a number of other nations. Many of the largest cities in the United States recovered from the hefty losses of the 1970s, and the growth of nonmetropolitan America petered out (Frey 1990).

What are the ramifications of these trends for megalopolis? Megalopolis emerged from the tendency for people to agglomerate into ever larger urban environments, and it was sustained by the powerful pull of centrality for transactional activities. During the 1970s the forces of agglomeration and the draw of centrality appeared to be fading. Mega-lopolis housed a dwindling share the nation's population, income, man-ufacturing activities, and corporate headquarters. Moreover, within Megalopolis, population and economic activity were being redistributed away from the central cores of its largest cities to suburbs and non-metropolitan fringes in response to individual preferences for lower-density living environments, the availability of cheap land for new construction, and technological innovations that made peripheral loca-tions more attractive industrial and office sites.

The renewed vigor of large metropolitan areas during the 1980s cast doubt on the counterurbanization thesis and implied that the counter-urbanization trends of the 1970s were merely a blip in the long-term trend toward ever greater agglomeration and concentration of the pop-ulation rather than another turning point in the history of human set-tlement (Champion 1989a). New York, Megalopolis's largest and most important city, emerged during the 1980s as a "world city" emphasiz-ing advanced services with international expertise in management, ad-vertising, law, accounting, and related activities. At the second echelon of the Megalopolitan urban hierarchy, Boston and Philadelphia be-came national command and control centers specializing in business and financial services, corporate decision making, and the distribution of consumer goods (Noyelle and Stanback 1984; Frey 1990). In mid-decade all three cities seemed poised to pilot the U.S. economy into the emerging global economy. The most recent recession of the early 1990s has not, however, been kind to the Northeast or to key Megalopolis cities, whose national and international command and control func-tions were battered by severe declines in white-collar employment—the hallmark of the so-called Yuppie recession (Gober 1993). Outmigration from the urban Northeast is again on the rise, sending demographers and urban geographers scurrying for new explanations of regional pop-ulation shifts.

The U.S. Bureau of the Census has been slow to adapt to the realities of megalopolitan-scale development. Indeed, until 1950 the census did not even officially recognize that cities and suburbs functioned to-gether as a unit. Only in 1950 did the census begin to publish statistics about standard metropolitan statistical areas (SMSAs) and urbanized areas (UAs). An SMSA is a "closely integrated economic and social unit with a large population nucleus, and generally consists of one or more entire counties that meet specified standards with regard to popula-tion, commuting patterns, and metropolitan character" (U.S. Bureau of the Census 1988:11). A UA also recognizes the inherent connection between city and suburbs and consists of at least one city with more than fifty thousand people and adjacent areas with a population density of at least one thousand persons per square mile.

Given the realities of large-scale urban development during the 1960s and 1970s, many formerly separate metropolitan areas merged (Dallas and Fort Worth are a prime example), and the expansion of the largest metropolitan complexes (Boston, for example) created separate metropolitan entities within the larger megalopolitan whole. Recognizing these developments, the Bureau of the Census in 1983 formally adopted the term consolidated metropolitan statistical areas (CMSAs), which are metropolitan complexes with more than one million persons created by the fusion of formerly separate SMSAs. In 1990 the United States had twenty CMSAs, led by the New York-Northern New Jersey-Long Island CMSA with eighteen million people, Los Angeles-Anaheim-Riverside CMSA with fifteen million, and Chicago-Gary-Lake County CMSA with a population of eight million (U.S. Bureau of the Census 1992).

Worried that the pace and nature of urban growth has outgrown even these hybrid conceptualizations of the nation's urban structure, the Bureau of the Census (in conjunction with the Office of Management and Budget) recently commissioned a group of geographers and demographers (Brian Berry, Richard Morrill, John Adams, William Frey, and Alden Speare) to develop new ways of representing the spatial structure of the settlement system. Although the results of the Metropolitan Concepts and Statistics Project are not yet finalized, the effort recognizes the increasing complexity of the settlement system, the importance of functional linkages in describing that system, and growing problems in using current standards. Urbanized areas that are increasingly bound up with, and within, one another are not well portrayed with current census geography.

The passage of time has introduced new terms for megalopolis's leading city—megacity, giant city, global city, and world city—but the ideas of bigness and transactional networks are recurrent themes. Interest in urban networks is especially strong in Europe because of the removal of internal boundaries and the opening up of Eastern Europe (Cambis and Fox 1992). Myriad city organizations have been established in recent years including Eurocities, a grouping of some of the continent's major "second cities." Eurocities want the European Community to facilitate relationships among them, contending that large cities are the engines for the community's future economic development. Urban success in the 1990s will be based on a city's ability to develop strategic alliances within Europe's city system. Key factors include the presence of multinational companies, centers of higher education and research and development, high-technology activities, communication infrastructures, business services, and a high quality of life (Borga 1992).

In a recent book, *The Global City: New York, London, Tokyo,* urban planner Saskia Sassen (1991) argues for the continuing importance of centrality in the international economy. Even as factories and production facilities decentralize to low-cost locations, financial trading, investment banking, and business services cluster in cities. Global cities have become the centers of the control and coordination for the global economy. Concentration of a new class of highly paid professionals in these cities and their reliance on low-wage service workers has created a new class structure in global cities. The global city is now the domain of the very rich and the very poor, the middle class having fled to the suburbs. The global city, according to Sassen, is the "brain" of megalopolis. New York City plays this role for the urbanized northeastern seaboard of the United States, Tokyo for the Tokaido Megaroporisu, and London for England's Southeast. These high-technology transactional spaces are the byproduct of the integration of the world's economy.

Many of these themes are present in geographer John Borchert's *Megalopolis: Washington, D.C. to Boston,* a field guide for the 1992 meetings of the International Geographical Union in Washington, D.C. Borchert updates many of Jean Gottmann's ideas about the essence of Megalopolis as a region, including the scale and density of urban life, the region's hinge function, the incredible intensity of traffic in both people and information within Megalopolis, and the effects of the information age on geographic patterns of people and economic activity with the region.

Even though almost all of what we know about megalopolis is based upon the experiences of more developed nations, it is clear that new megacities are emerging in less developed countries. According to UN projections for the year 2000, twenty-three of the world's megacities (each with more than eight million inhabitants) in the year 2000 will be in developing countries. The combined population of the São Paulo-Rio de Janeiro conurbation will near thirty-five million; Bombay alone is projected to house eighteen million and Lagos over thirteen million (UN Dept. of Economic and Social Information and Policy Analysis 1993). The sheer size of these agglomerations necessitates a polycentric urban structure and ensures that they will bring the surrounding countryside into their spheres of influence in much the same way that the original Megalopolis came to dominate the space economy of the United States.

Less clear is whether the megacities of the Third World will replicate megalopolis's specialization in information processing and advanced services and its role as an economic hinge effectively linking the national and international economies. Size and physical structure alone do not add up to Gottmann's vision of megalopolis as the cradle of civilization and brain of the modern economy. Some of the economic advantages of large markets and economies of scale in Third World megacities must be counterbalanced by the disamenities of large-scale urban agglomeration such as poor air and water quality, chronic traffic congestion, inadequate solid waste disposal, sewerage deficiencies, and high crime rates (Richardson 1993). The tradeoffs between the economic advantages of extremely large-scale urban development and attendant environmental problems in some of the world's poorest countries, such as India (Bombay, Calcutta, and Delhi), China (Beijing, Shanghai, and Tianjin), Bangladesh (Dacca), Pakistan (Karachi), and Nigeria (Lagos), are only beginning to be realized.

Conclusions

The idea of megalopolis and the intellectual activity derived from it ignited scholarly and public discourse about cities and urban growth. The result has been a growing awareness of the enlarged scale of modern urban development; the geographic outcomes of the shift from manufacturing to services; the effects of the global economy on the size, growth, and function of cities; the emergence of transactional spaces and the effect of such spaces on the lives of people. In addition to influencing geographers, planners, and architects, megalopolis infiltrated the world view of educated people and served as a vehicle for interpreting the momentous economic, social, political, and geographic changes taking place in cities in the late-twentieth century.

Generalizations about megalopolis seem obvious to us today. Who would question the existence of functionally interconnected systems of cities, the inevitability of widespread suburbanization, the emergence of a global economy, or the importance of transactions? But, these processes were hardly recognized when Megalopolis was first described and its essential features generalized to other places.

Equally significant is that megalopolis was spawned from a discipline that historically had had little interest in cities and urbanization, even though a majority of the nation's population lived in cities. Today urban geography is the second largest of the Association of American Geographers' specialty groups with 637 members (Geographic Information Systems is the largest with 1,007).

In conceiving and popularizing megalopolis, Jean Gottmann provided the discipline of geography and the broader field of urban studies with a new and rich set of ideas about the way humans organized themselves in space and shaped the environment of their lives. That megalopolis today is far more frequently cited outside than inside geography and used more often by journalists and lay people than by academics is a testament to the ability of this idea to penetrate both disciplinary and academic boundaries.

References

Beale, C. L. 1975. *The Revival of Population Growth in Non-metropolitan America.* Economic Research Service, ERS 605. Washington, D.C.: U.S. Department of Agriculture.

Berry, B.J.L. 1976. The Counterurbanization Process: Urban America since 1970. In *Urbanization and Counterurbanization,* ed. B.J.L. Berry, 17–30. Beverly Hills: Sage.

Bigham, T. C., and J. M. Roberts. 1962. *Transportation, Principles and Practice.* New York: McGraw-Hill.

Boorstin, D. 1961. *The Image.* New York: Atheneum.

Borchert, J. R. 1992. *Megalopolis: Washington, D.C. To Boston.* New Brunswick: Rutgers University Press.

Borga, J. 1992. Eurocities—A System of Major Urban Centers in Europe. *Ekistics* 59:21–27.

Cambis, M., and S. Fox. 1992. The European Community as a Catalyst for European Urban Networks. *Ekistics* 59:4–6.

Champion, A. G. 1989a. Counterurbanization: The Conceptual and Methodological Challenge. In *Counterurbanization,* ed. A. G. Champion, 19–33. New York: Edward Arnold.

———. 1989b. Introduction: The Counterurbanization Experience. In *Counterurbanization,* ed. A. G. Champion, 1–18. New York: Edward Arnold.

Currie, L. 1976. *Taming the Megalopolis.* New York: Pergamon Press.

Doxiadis, C. A. 1974. *Ecumenopolis: the Inevitable City of the Future.* New York: Norton.

Frey, W. H. 1990. Metropolitan America: Beyond the Transition. *Population Bulletin* 45:1–51.

Gober, P. 1993. Americans on the Move. *Population Bulletin* 48:1–39.

Gottmann, J. 1961. *Megalopolis: The Urbanized Northeastern Seaboard of the United States.* Cambridge, Mass.: MIT Press.

———. 1970. The Green Areas of Megalopolis. In *Challenge for Survival: Land, Air, and Water for Man in Megalopolis,* ed. P. Dansereau, 61–65. New York: Cambridge University Press.

———. 1976. Megalopolitan Systems Around the World. *Ekistics* 41:109–113.

———. 1983. Urban Settlements and Telecommunications. *Ekistics* 48:411–416.

Hall, P. A. 1971. *London 2000.* London: Faber and Faber.

Hanes, J. E. 1993. From Megalopolis to Megaroporisu. *Journal of Urban History* 19:56–94.

Isomura, E. 1969. *Megalopolis in Japan: Its Reality and Future.* Tokyo: Nihon Keizai Shimbunsha.

Jacobs, J. 1961. *The Death and Life of Great American Cities.* New York: Random House.

Leman, A. B., and I. A. Leman. 1976. *Great Lakes Megalopolis: From Civilization to Ecumenization.* Ottawa, Canada: Ministry of Supply and Services.

Marston, S. A., G. Towers, M. Cadwallader, and A. Kirby. 1989. The Urban Problemmatic. In *Geography in America,* ed. G. Gaile and C. J. Wilmott, 651–672. Columbus: Merrill.

Mumford, L. 1961. *The City in History.* New York: Penguin Books.

Noyelle, T. J, and T. M. Stanback. 1984. *The Economic Transformation of American Cities.* Totowa, N.J.: Rowman and Allanheld.

Pred, A. R. 1966. *The Spatial Dynamics of U.S. Urban-industrial Growth, 1800–1914.* Cambridge: MIT Press.

Richardson, H. W. 1993. Efficiency and Welfare in LDC Megacities. In *Third World Cities,* ed. J. D. Kasarda and A. M. Parnell, 32–57. London: Sage.

Sassen, S. 1991. *The Global City: New York, London, Tokyo.* Princeton, N.J.: Princeton University Press.

Solot, M. 1986. Carl Sauer and Cultural Evolution. *Annals of the Association of American Geographers* 76:508–520.

Swatridge, L. A. 1971. *The Bosnywash Megalopolis.* Toronto: McGraw-Hill.

Taaffe, E. J., and H. L. Gauthier. 1973. *Geography of Transportation.* Englewood Cliffs, N.J.: Prentice-Hall.

U.N. Department of Economic and Social Information and Policy Analysis. 1993. *World Urbanization Prospects: the 1992 Revision.* ST/ESA/SER. A/136. New York: United Nations.

U.S. Bureau of the Census. 1975. *Historical Statistics of the United States.* Washington, D.C.: U.S. Government Printing Office.

———. 1988. History and Organization. In *Factfinder for the Nation,* CCF No. 4. Washington, D.C.: U.S. Government Printing Office.

———. 1992. *Statistical Abstract of the United States: 1992.* Washington D.C.: U.S. Government Printing Office.

PART 4

10

Sense of Place

Edward Relph

Ignorance of Place, or King Lear in Las Vegas

I was on the main strip in Las Vegas, enjoying the regularly scheduled volcanic eruption in the tropical gardens of the Mirage Hotel, when a remark of King Lear's sprang unheralded to my mind: "I am mainly ignorant what place this is." A simultaneous rift opens in fiction and physics, and a confused old king is transported through time, space, and reality. The sidewalks are crowded with pedestrians, but they are mostly intent on their own fortunes or misfortunes as, clutching plastic cups of quarters for slot machines, they move between casinos and Lear attracts little attention. In the play he was, by his own admission, "not in perfect mind." The world around him was sane enough; his geographical problem was in his head. Here the opposite seems to be the case.

The streetscape of the strip is one of unpredictably juxtaposed scenes: seven lanes of traffic, signs three stories high, a Hindu shrine, buildings covered in decorations suggesting a Mississippi riverboat, an Irish pub, something Japanese, and countless other fancies, and, behind all of this, the modernist slabs of skyscraper hotels. Snatches of tropical bird songs come from speakers, carefully hidden behind signs advertising the time of the next volcanic eruption. A few yards away two full-size sailing ships engage in mock battle in front of a vaguely Caribbean village. When the last pirate has fallen from the rigging, Lear, his own distractions distracted by these marvelous scenes, wanders off into a fragment of rain forest with a complex synthetic ecology of imported and fiberglass palm trees.

Public and private spaces here blend seamlessly into one another. Lear unthinkingly drifts out to the street again, then is attracted through a gilded classical portico and onto a moving sidewalk, which transports him regally to the Appian Way, part of the Forum Shops at Caesar's Palace. The architecture is Roman, the designer stores are Italian, and the crowds watching imagineered statues of Bacchus and lesser deities as they burst into a regularly scheduled, animatronic laser show are mostly groups of Japanese tourists. A brilliant azure sky with slowly moving cumulus clouds is projected onto the domed ceiling until whatever time the mall closes, and then it fades through sunset to a starry canopy.

Along the street a temporarily abandoned lot, where the Dunes Hotel once stood and will stand again, has begun to revert to desert, and trash has blown up against the surrounding chain link fence. Beyond it is a preposterously impossible medieval castle, with roofs of bright red and blue, exaggerated towers, and fanciful fortifications. A monorail links this edifice to an immense glass pyramid surrounded by sphinxes, obelisks, holographic sculptures, and palm trees shaped in the ancient Egyptian manner (Figure 10.1). From its peak a laser light pierces the heavens with such intensity that whenever a mutual thinning of smog permits, it can be seen in Los Angeles. Across the street a cluster of Easter Island heads stare with disdain at the passing traffic and at a huge supine lion embedded in a massive block of a dark glass building. The lion's mouth is the entrance to the MGM Grand, with its 5,000 room hotel and theme park. Inside, statues of Dorothy, Toto, the Tin Man, the Lion, and the Scarecrow dance forever along the yellow brick road away from Oz and toward the casino.

Like Dorothy and her friends, King Lear is trapped in Las Vegas in a single dramatic moment. He is muttering repeatedly to himself: "I am mainly ignorant what place this is." His ignorance is understandable. Clearly nobody can be entirely sure what place Las Vegas is. It is tempting to assume that beneath the decorations and signs there is no here here, nothing intrinsic except a patch of semi-desert with chain link fences and wind-blown garbage. This assumption should not be made too quickly. If we look carefully we will see little pretense, and we will discern no real attempt to disguise the borrowing and contrivance. It is hard to imagine anyone being fooled into thinking that this is a real volcano or a real medieval castle. In this sense, almost everything here is an honest fake.

The Las Vegas strip is a remarkable conflation of the best bits of

Figure 10.1. Tourists in Egypt in Las Vegas: The Luxor Hotel. Sketch by author.

geography and history, reconstituted and recombined. It is a place made of other places and other times, of fantasies where the real and the artificial slide easily into one another, created to house a dream of unlimited freedom through instant wealth. The designers of Las Vegas are not misguided; they have created somewhere that now attracts twenty-three million visits a year. And the tourists who make those visits are not deluded; they know all this is fabricated. They come because they like Las Vegas. For me, as a geographer, it is largely immaterial whether or not I agree with them. The real challenge is to get some critical understanding of what is happening here, or, in other words, to make sense of this place.

Sense of Place

Sense of place is not a theory that geographers invented and now offer as a service to the rest of society, nor is it a tool in the way that a map or central place theory are tools that unlock geographical puzzles. Shakespeare needed to know nothing of geography to grasp the importance of sense of place if King Lear was to get his bearings, and tourists in Las Vegas need to know nothing of geography either to get there (tour packages do that) or to enjoy the juxtaposed place fragments (for which they have presumably been prepared by a lifetime of watching discontinuities on television). Sense of place is first of all an innate faculty, possessed in some degree by everyone, that connects us to the world. It is an integral part of all our environmental experiences and it is only because we are first in places that we can then develop abstract arguments about environment, economy, or politics. But in addition to this, sense of place can be a learned skill for critical environmental awareness that is used to grasp what the world is like and how it is changing. Geographers have reflected on sense of place as a faculty, and they have developed it as a skill, throughout the history of their discipline. They have often done this through writing, but they have done it more frequently, I believe, through their teaching, passing both the understanding and the skill of sense of place like cherished traditions from one generation to the next. It would be illogical to claim that sense of place, which has to do with specific contexts, has led to some universal change in environmental knowledge and practice. What it has done, and should continue to do, is contribute common sense and understanding to countless local changes to the world.

It is not only geographers who have attended to sense of place. Architects, psychologists, psychiatrists, artists, literary critics, poets, and even, though in my opinion with little success, economists, have all considered aspects of it. Most of them would have us believe that sense of place is invariably good, and that its enhancement can only make built environments more beautiful, our lives better, and communities more just. The geographical view is broader and less idealistic. For geographers, places are aspects of human life that carry with them all the hopes, accomplishments, ambiguities, and even horrors of existence. They see sense of place as a thread that ties each of us to our surroundings, and as a learned way for understanding somewhere on its own terms. As a form of environmental connection, sense of place is existential and political. As a learned geographical skill, it requires careful and critical observations of places and the recognition that surface appearances can reveal shared cultural traditions or disguise profound injustices. Indeed, the political desire to belong to some place and to participate in its traditions can fuel attempts at the systematic exclusion of all those who are believed not to belong. In contrast to those who may believe that a stronger sense of place can only make the world better, geographers try to remember that a persistent solution to social problems has been to destroy the places and communities where these occur, perhaps under the name of urban renewal or resettlement.

It is a perverse testament to the importance of sense of place that one of the most enduring of all military strategies is the destruction of places in order to undermine the will of a people. It is, of course, an even more powerful testament to the importance of sense of place that these places are invariably rebuilt by those whose attachment to them has not been destroyed.

In short, sense of place is a strong and usually positive faculty that links us to the world, but it can also become poisoned and destructive. As a taught skill, geographical sense of place has always aimed to grasp both what is good and what is bad in places, then to argue critically for changes that are just and enduring, yet responsive to diverse environments and cultures. This skill was first described by geographers over two thousand years ago. It was considered important then; it is essential now if we are to unravel the increasingly confused geographies of the late-twentieth century.

A History of Place in Geography

In antiquity it was a common practice for people to identify themselves by their name and the place from which they came, a geographical tradition, perhaps begun by Herodotus, which has sadly lapsed. Eratosthenes of Cyrene, a town on the coast of what is now Libya, was for some years around 225 B.C. the librarian at Alexandria and is the person usually credited with inventing the word and the idea of "geography." Not much is known about him. Apparently he was a person of many talents—a poet, mathematician, and historian as well as the first geographer. I imagine him to have been a lean and serious man, with a librarian's propensity to classify. He invented geography, I suspect, not in a moment of epistemological inspiration but out of desperation in order to organize all the information about different places he had gathered from reports of traders and travelers. His three books of geography, now lost, apparently consisted of an account of a measurement of the size of the earth, a map of the known earth, and a set of descriptions of countries.

From the little we know, it seems that for Eratosthenes sense of place consisted mostly of accurate knowledge of the locations of places. It was a far more complex notion for Strabo of Amasia, a town on the south coast of the Black Sea, who is perhaps the most famous geographer of antiquity. His multi-volume book, *Geography*, probably written between 9 and 5 B.C., is one of the oldest extant works of the discipline. Strabo acknowledged Eratosthenes (and is in fact the source of much that is known about him), yet dismissed contemptuously both his cartographic approach and his claim to be the founder of the discipline. Instead Strabo, in accord with his beliefs as a Stoic, traced the origins of geography to Homer's insight that the earth and all its places had been created by the gods for the use of humans. This god-given geography had a providential order; some regions were endowed with good properties for human activities and other regions were poor. Humans must use their powers of foresight and reason, also provided for them by the gods, in order to recognize and respond to this providential order; if they fail to use reason they will probably end up eking out a miserable existence on some barren heath.

"A knowledge of places," Strabo wrote at the beginning of his *Geography* (section 1.2.12), "is conducive of virtue," and virtue means living in harmony with nature (Strabo 1917). Geographers, with their knowledge of the cosmos, wide travel, and careful observations, were especially capable of evaluating places and the ways of life of people. This acute sense of place enabled geographers to interpret the providential order of landscapes and to distinguish good settings from unpropitious ones. They could also advise others on how to recognize or achieve for themselves the Stoic ideal of living virtuously, and accordingly Strabo directed his book especially to political and military leaders.

I admire Strabo's view of geography, and my interpretation is probably biased by this admiration. His *Geography* is not a classical foundation for environmental ethics in the 1990s, as my account would seem to imply. His many-volume book is mostly an uncritical list of details that has been described as "an atlas in prose"; some of the information in it is quite fantastic and not at all reasonable or grounded in observation. Nevertheless, it is no misrepresentation to suggest that Strabo believed that geographers have a special sensitivity to the qualities of specific places and that their training allowed them to see through the surface forms of landscapes to a subtle and divine order. Strabo's understanding was therefore in marked contrast to that of Eratosthenes, for whom place was factual and objective, largely a matter of location and shared properties.

This difference in interpretation of place is one that has echoed through the discipline of geography ever since. Since Strabo, geographers have often written carefully observed and thoughtful accounts of places but have written little about the idea or sense of place. For those working in the tradition of Eratosthenes, the objective qualities of places, especially their location, have mattered most. For other geographers the uniqueness and virtue of places have been more important. Nicholas Entrikin (1991), a geographer from Los Angeles, has recently tried to embrace this ambivalence by describing it as "the betweenness of place"—meaning that sense of place sits somehow between the objectively shared properties of environments and subjectively idiosyncratic experiences of them. From the objective vantage point, place is regarded either as location or as a set of shared relationships; from the subjective perspective, place is a territory of meanings and symbols. "To understand place," Entrikin (1991:5) has written, "requires that we have access to both an objective and a subjective reality. . . . Place is best viewed from points in between." In other words, with something borrowed from both Eratosthenes and Strabo.

I think this is too theoretical. Observation of somewhere always reveals shared or borrowed elements (such as the pyramid, Easter Island statues, and imported palm trees in Las Vegas), which can appropriately be described as displaced or placeless, and intrinsic or distinctive characteristics (such as Caesar's Palace or the one-of-a-kind strip in Las Vegas). The placeless bits are the outcome of general, and probably objectively developed, processes, yet they have been incorporated into a specific context. We experience a world of subjective specifics, not one of objective generalizations, and it is not possible to situate oneself self-consciously at points between what is objective and subjective. Enquiries into place should begin with specifics, then through those explore the interesting questions about how the intrinsic and the placeless aspects fit together, and in what sort of balance.

Here's an example taken from John Ruskin, the nineteenth-century critic of art and of industrial society. In the preface to his book *The Crown of Wild Olive*, published in 1866, he describes a stream at Carshalton pools in the south part of London, once clear and unsullied, but by then overwhelmed by development, where "the human wretches of the place cast their street and house foulness; heaps of dust and slime, and broken shreds of old metal, and rags of putrid clothes" (Ruskin 1866:386). Half a dozen men, in one day's work, could clean it up, but, Ruskin comments astutely, that day's work is never done. He walked up the hill from the stream and came upon a new tavern, in the front wall of which there was a recess about two feet deep that was fenced off with an imposing iron railing serving no purpose except to protect the refuse blown behind it.

> Now the iron bars that, uselessly, enclosed this bit of ground, and made it pestilent, represented a quantity of work that would have cleansed the Carshalton pools three times over;—of work, partly cramped and deadly, in the mine; partly fierce and exhaustive, at the furnace; partly foolish and sedentary, of ill-taught students making bad designs. . . . Now how did it come to pass that this work was done instead of the other; that strength and life were

spent in defiling the ground instead of redeeming it; and in producing an entirely (in that place) valueless piece of metal, which can be neither eaten nor breathed, instead of medicinal fresh air and pure water? (Ruskin 1866: 387–388)

A pointed question, one that could be asked of many places now. It led Ruskin into a blistering critique of nineteenth-century industrial economies.

Ruskin's description sees no issue about betweenness, no problem of oscillation between the objective and subjective aspects of Carshalton. Rather, he discovered *in* the specific characteristics of this place the embedded significance of widespread industrial practices. Abstract process is revealed through the particularities of the place. A well-developed geographical sense of place is one that looks carefully at local idiosyncrasies, keeps an open mind about them, and then sees *through* these to the larger patterns and processes they signify. The inverse moral to this is no less important. Social theories and abstractions, for instance about progress or economic growth, have substance only in the actual lives of individuals in particular places.

The Disappearance of a Perfect Sense of Place

Old psychology texts often included a diagram of a homunculus, a creature the shape of which is determined by how much of the brain is devoted to sensations from various regions of the body. This is not an attractive thing. It has a tiny body, short limbs, a huge face with large lips, big feet, and bigger hands. The sense of sight is located in the *area striata*, substantial sections in both halves of the brain, and, if these are included, the homunculus also has monstrous bulging eyes.

Some of our senses have organs that conduct information from the world to our brains, and others do not. Unfortunately the homunculus is missing both an organ and a cortical region for the sense of place. An extra organ for this—another nose, perhaps, or a third eye—would no doubt clarify environmental relations enormously. As it is, all we can say is that sense of place is a synthetic faculty, unifying the information reaped by other senses. It is probably best thought of as a web having no fixed location in the brain. But one thing about it is clear: it overlaps extensively with the part of the memory reserved for nostalgia and golden ages because almost everything written about sense of place extols what is old or traditional and decries whatever is new.

The age of perfect places is not fixed in history. For some, such as the social psychiatrist Eric Walter (1988), that age was the classical period when the worlds of the gods and humans apparently coincided. For others it lies in the Middle Ages or the Renaissance; for Tony Hiss, a journalist from New York who has written on *The Experience of Place,* it seems to be any time in the northeast United States before about 1930. More elusively, Christopher Alexander, an architect originally from England now living in California, has discovered it in any setting that possesses "a quality without a name"—his term for good properties of places we can recognize well enough but somehow cannot precisely define. The illustrative photos in his book *The Timeless Way of Building* include many vernacular settings, Greek and English villages, and fragments of central Paris and Amsterdam, and he writes specifically of a Japanese farm where the carp swim in a pond as though for all eternity. Alexander's (1979:164) argument is that wherever love, care, and patience are in adjustment with environment, then "human variety, and the reality of specific human lives, can find their way into the structure of the places."

The usual explanation for why old places were so much better than recently built ones can be found, for example, in an argument made by Michael Ignatieff (1984:138), a cosmopolitan (his self-description) Canadian philosopher and journalist living in Europe, in his book *The Needs of Strangers.* Until the beginning of the present century, he writes, most people's lives were bounded by the distance they could walk or ride in a day. Local dialects and identities were strongly marked and were reinforced by building styles based on the use of local materials and regional traditions. A concordance of social values, technologies, and environment prevailed, a concordance expressed in an ancient language about roots, spirits of place, and the need to belong somewhere. Sense of place was indeed a powerful and positive force.

In 1913, Vidal de la Blache, a prominent French geographer who took great delight in the diverse landscapes of his home country, wrote an essay on the character of geography in that he defined the discipline as "the study of places" (Vidal de la Blache 1913). The landscapes that attracted Vidal's attention were filled with a diversity of places, relatively shielded from outside influences by distance and by the resilience of local cultures. It seemed entirely appropriate that geographers, with their long-standing concern for the appearance of the world, should be dedicated to making sense of this diversity. It was, of course, recognized that the boundaries of real places were permeable. Travelers, pilgrims, scholars, and itinerant artisans brought with them knowledge from elsewhere, but that knowledge was invariably adapted to local traditions, building styles, and narratives, rather than being imposed on them.

There seems to be good reason to believe that in the premodern world the local ways of doing things rested in a fine balance with imported, universal practices, and the result was an intelligible diversity of landscapes—every place was distinctive yet not so different that it was incomprehensible to outsiders.

Theoretical knowledge sometimes has such force that it leads to substantial changes in the social order and the ways people live. This could be claimed, for example, of Newton's physics and the various philosophies of the Age of Reason: first these were thought, and then the world changed to conform more closely to their image. At other times, the world itself changes and theoretical knowledge struggles to keep pace, to find ideas and images that will explain what is happening. Between 1850 and 1950, and to some extent even now, geographical understanding of places has struggled to keep pace with reality. Even as Vidal wrote in 1913, his definition of geography was obsolescent. For decades the resilience of local culture had been under assault from technological and political processes. In the entry under "Geography," the Oxford English Dictionary provides two revealing quotations from the mid-nineteenth century that illustrate this well. "We have seen the railroad and the telegraph subdue our enormous geography," wrote Emerson in 1854; in 1859 Lever declared that "Science has been popularized, remote geographies made familiar." Even then there was a feeling that a new logic of geography was emerging, one in that the distinctiveness of places would be suppressed. Walt Whitman, ever attuned to what the future was bringing, caught a whiff of this change, ironically in his precisely located poem "Crossing Brooklyn Ferry" in *Leaves of Grass* (1855): "It avails not," he wrote, "time nor place—distance avails not."

In this matter geographers were hopelessly out of touch with their own subject. In their research and writing they mostly ignored cities and industrial economies and instead continued to study regions unaffected by these dramatic changes. It was almost as though most of them had come to an agreement to maintain a comfortable belief in the importance of regional diversity regardless of the evidence. It was, in fact, not until the 1950s that geographers began to write about the new economic and urban processes (as Pat Gober notes in Chapter 9 of this book). I do not believe it is an exaggeration to call this a long period of collective blindness. It was as though an entire discipline had been restrained in a Plato's cave of its own devising, determined to seek out regional variations and place distinctiveness no matter what. It is a salutary lesson of how easy it is to ignore the evidence of one's own senses, including the sense of place.

Modernism and Placelessness

Premodern places looked as they did largely because of their geographical and cultural context and their relative remoteness. Their appearances were as much as matter of necessity as anything else, and little about living in them was romantic. I can attest to this because I grew up in a village in South Wales overlooking the Wye Valley—barely a village, more a scattering of houses—that had no running water or electricity until the mid-1950s. In part because of its backwardness, this was a strongly independent community; everyone knew and was known by everyone else, and many people lived their entire lives never traveling more than a few miles away. The village possessed remarkable resilience in the face of hardships, such as winters when the roads were blocked for weeks by snow. But for all the positive qualities this was not a particularly comfortable or convenient place to live, and in the 1970s, when life in the countryside became attractive for middle-class people from cities, many of the local residents jumped at the chance to sell their property and moved to nearby towns. Their damp little cottages were deeply renovated by the newcomers, or replaced by neat subdivisions of big houses with suburban sidewalks and street lights. The new residents commute long distances to work (some to London, more than 100 miles away), take their holidays in Florida or Turkey, have revived moribund festivals and created a new community life. One old pub has been converted into a French restaurant with a clientele that includes Hollywood movie stars. The village is in the same location where I grew up, but it really is a different place.

I do not see anything remarkable in the changes that have occurred here, except perhaps the relative lateness of their arrival. Since 1900 similar geographical and social changes have permeated villages, urban neighborhoods, and towns around the world. These changes, which have come in two waves—modernism, then postmodernism—have profoundly altered the appearance and the meanings of places.

In the decade at the beginning of the twentieth century, when Vidal was celebrating geography as the study of places, groups of artists, poets, and architects across Europe were simultaneously struggling to cast off the baggage of tradition and to reinvent society and art along lines that reflected the new technologies of electricity, automobiles, and mass production. They looked to the future and not to the past for their inspiration. The results were dramatic and unprecedented: the abstract paintings of Braque and Picasso, the dancing of Duncan and Nijinsky, and the ascetic geometric and unornamented buildings of Gropius, Le Corbusier, and the Bauhaus. This was modernism. It was a grand social theory and it allowed no space for tradition, convention, decoration, or local culture.

If there was an early nerve center for modernism, it was the Bauhaus, a design school based at Dessau in the eastern part of Germany in the 1920s. The Bauhaus members were an eclectic group of artists and architects who developed unornamented, streamlined, and geometric designs for everything—chairs, typefaces, fabrics, electric light fixtures, kitchen appliances, houses, offices, factories, city plans, and their own school buildings. The aesthetic principle behind these designs was that they should appear to be functional and futuristic (though anybody who has tried to sit in Bauhaus chairs knows that function was sometimes only in the appearance). The social ideology behind them was democratic—they should be capable of being mass-produced and thus available for everyone. Houses, said Walter Gropius (1965:39–40), the architect who was the director of the Bauhaus for many years, should be mass-produced in factories, and every part of the house should therefore be standardized. Distinctiveness would be the result of individual expression using these standard parts, not historical style or locality. In other words, modernist designs have no need for geography; they are equally applicable anywhere.

This proved to be a valuable precept. When the Bauhaus was closed by the Nazis in the early 1930s, its architect members scattered according to their political inclinations either to North America or to the Soviet Union, where they could continue to do much as they had in Dessau; in the former they came to specialize in glassy skyscraper offices, and in the latter they designed geometrically arranged suburbs of apartment slabs. In the 1950s and 1960s, when cities in Europe and North America were expanding rapidly or being radically renewed, Bauhaus architects and their modernist disciples were conveniently available with their placeless designs. These fit so well with the intentions of international business that the two soon became almost indistinguishable. Holiday Inns, McDonald's, Sony, IBM, Volkswagen, and Shell have standardized buildings or products—everywhere. If reconstructed city centers and newly constructed suburbs retained some distinctiveness it was usually because of old road patterns or names; the components of them—such as office buildings or franchises—were often identical regardless of location. The balance between the local and the universal had been shifted, and sameness had begun to overwhelm geographical difference.

As they awakened in the 1950s from the discipline's long doze, a group of geographers noticed this emerging uniformity and celebrated it in theories of uniform space and central places (see Chapter 8 of this book). In the 1970s a second group began to look differently and saw little reason to celebrate. For them it was apparent that the diversity of places so long revered in their discipline was being systematically eradicated by a modernist consortium of architects, planners, and international business, with the help of some of their colleagues in geography. They protested, not loudly and not stridently, but thoughtfully and deliberately. The focus of this intellectual protest was Yi-fu Tuan's widely read book *Topophilia* (that literally means love of place). In the introduction Tuan was explicit that his interest was not with applied knowledge that would change the world, but with how we might better understand ourselves by understanding our environmental attitudes. In *Topophilia* he wrote of place and diversity as aspects of positive environmental experiences. He wrote of home and nostalgia, of utopias, of personal experience, cosmos, and symbolism. He wrote of ideal places and of environments of persistent appeal such as the seashore, the valley, the island, the wilderness and the mountains. He wrote of matters that many of his modernist colleagues thought were extinct. In fact, it turned out that he was gently announcing the next wave of change.

Postmodern Place Revival

In 1968 the architect Robert Venturi took a group of graduate students from Yale University to Las Vegas to investigate the commercial strip there. Modernists, insofar as they were able to bring their attention to bear on it at all, could only condemn the profusion of lights, signs, architectural fragments, and aimless spaces; to them it was dysfunctional architectural sewage. What Venturi and his students saw, however, was a wonderful vitality of architectural styles, decorations, heraldic signs, and ceremonial spaces. This was a way of seeing they had learned in part from J. B. Jackson, perhaps best described as a geographer by choice, who had been writing eloquently about ordinary American landscapes and sense of place for almost half a century (see, for example, Jackson 1970).

Learning from Las Vegas (Venturi, Scott-Brown, and Izenour 1972) punctured the modernist bubble in architecture. Since then, buildings of all types have been increasingly decorated with colored trim, peaked roofs, Romanesque arches, and neoclassical columns—sometimes all at once. This self-consciously historical and decorative approach has come to be called the postmodernist style.

Postmodernism is more than an architectural fashion. It has counterparts in literature, art, and philosophy, and the term has been muddied by obscure academic debate. The idea is nevertheless clear enough; it describes something that simultaneously comes after and takes issue with the major tenets of modernism. Modernism stood for the future, for standardization, for undecorated functionality; post-modernism celebrates the past, difference, decoration, and unpredictability. It does this with an affected sense of irony because it exists only in relation to modernism, and underneath their decorative facades postmodern buildings are decidedly high tech, with steel frames, air conditioning, fiber optic cables, and talking elevators.

Some origins of postmodernism seems to lie in the various protest movements of the 1960s—civil rights, anti-war demonstrations, the women's movement, and environmentalism. Jane Jacobs's *The Death and Life of Great American Cities* (1961) was a polemic against modernism and a plea for protecting urban neighborhoods, and it was in neighborhoods that many protests against urban renewal and expressways erupted. Matters of history, which in architecture and planning had been suppressed for several decades, were now suddenly rediscovered and given the shield of heritage to protect them from further threats. Until about 1965, heritage concerns were limited to the protection of a few politically significant sites such the birthplaces of presidents; in 1968 there were no entries in the National Register of Historic Places in the United States, but by 1978 there were almost twenty thousand. This discovery of history was accompanied by a renewed enthusiasm for geographical diversity. Distinctive places had never entirely lost their appeal as tourist attractions, though in cities in the 1950s many were willfully destroyed by urban renewal or suburbanization. After about 1970, anywhere with a picturesque townscape, good scenery, pleasant climate, sand beaches, or ideally all of these together, became desirable as a place to live or as a tourist attraction. The village in which I once lived in South Wales was designated as an Area of Outstanding Natural Beauty and attracted a new breed of residents; towns in Provence that had almost been abandoned as inefficient backwaters were renovated as communities of second homes for Parisians or the British; and inner-city neighborhoods of decaying houses in New York, Philadelphia, San Francisco, and Toronto were gentrified by young professionals.

This revival of sense of place is not without its difficulties. The condition of postmodernity, as geographer David Harvey (1989) has called it, has become aligned with increasingly subtle forms of exploitation even as it apparently celebrates differences. He comments that sense of place in a postmodern world is exploitable for profit, and writes, "the search for roots ends up at worst being produced and marketed as an image. . . . At best, historical tradition is reorganized as a museum culture . . . of local history" (Harvey 1989:303). Blist's Hill Museum at Ironbridge in Britain, the supposed hearth of the Industrial Revolution, is made up of authentic industrial buildings that have been trucked in to create a town that never existed, staffed by the otherwise unemployed of the late twentieth century dressed in period costume to act like the workers of the mid-nineteenth century. Hundreds, perhaps thousands, of such historical settlements have been created around the world. Most of them do, at least, have some relation to local history and geography, albeit moved and sanitized. But this connection can be stretched a very long way, and is not essential. London Bridge has been moved to Arizona, and gondolas ply the waterfront in Toronto.

The message behind all this is simple: people prefer distinctiveness, therefore distinctiveness should be created. In this creative process geographical context is a useful resource but no constraint. Any interesting locality with strong popular appeal will do. An important question now, the anthropologist Clifford Geertz (1988:131) has written, is "what happens to reality when it is shipped abroad?" The simple answer to this is theme parks, which replicate and idealize otherwise remote environments and places so that visitors can enjoy the best of everywhere without the hardships of travel. A slightly more complicated answer is manifest in the strip in Las Vegas, which is being transformed into a sort of theme park without walls, a walk-in hologram, a virtual geography of the best place and time images from ancient Egypt, Polynesia, medieval England, the tropics, ancient Rome, or anywhere that appeals to the imagination. The strip is, paradoxically, a distinctive place comprised largely of fanciful fragments of other places. Similar geographical confusions, albeit on a less sophisticated scale, can be found in the mundane settings of the food courts of shopping malls, with their array of international fast food franchises, or in new housing developments that freely borrow their street names and building styles from around the world. We live, Clifford Geertz (1986: 121–122) observes, more and more in an enormous collage, among migrations of cuisine, of peoples, and of architectural styles.

I suppose the process of replication and borrowing from elsewhere could be considered geographical quotation, or, less charitably, place plagiarism. Whether it is reprehensible or a source of enjoyment is not as important to me as the fact that a new logic of place and geography is at work here. The premodern logic was that place identity grew from the location and its traditions, and this revealed itself in geographical diversity. Customs and styles did move from region to region but these processes of borrowing were relatively subservient to local distinctiveness. The modernist logic was that place was irrelevant and geo-

Figure 10.2. "We have leased these places to stand to private corporations" (McLuhan 1964:73). Coca-Cola celebrating Montreal in a bus shelter sign. Photo by author.

graphies should be determined by international economic forces and fashions; the manifestation of this was placelessness in that locality was subservient and places came to look increasingly alike. The postmodern logic of places is that they can look like anywhere developers and designers want them to, and in practice this is usually a function of market research about what will attract consumers and what will sell. In postmodernity it is as though the best aspects of distinctive places have been genetically enhanced, then uprooted and topologically rearranged. Marshall McLuhan, the media guru from Toronto, anticipated this sequence. In *Understanding Media* (1964:73) he quotes Archimedes—give me a place to stand and I will change the world—then remarks caustically, "we have leased these places to stand to private corporations" (Figure 10.2).

Sense of Place versus Geographical Gullibility

Much of what is manifest in postmodern places is geographically irreverent. It plays fast and loose with context and identity. It breaks long-established conventions about what places are and where they can be. The results, as in Las Vegas, can be seen as superficial and commercial, but they are also often self-consciously amusing, and I think we have to take care not to become too pompous and self-righteous in judging them. On the other hand it is important not to be too accepting and too gullible about what is going on because so much in postmodernism is about deceptions: concrete made to look like dressed stone, fiberglass made to look like palm trees, new buildings made to look like old buildings, and neotraditional new towns with quaint street patterns and regional styles of architecture developed by global corporations. It is important to know what belongs, what has been imported, and what has been invented in a place. It is important, in other words, not to be fooled by what is going on.

A major task in teaching a geographical sense of place now is to convey what might be called cheerful suspicion. This involves careful, unprejudiced observation of places and landscapes that is neither supercilious nor cynical. It requires that we sort out the elements of a place, how these relate, and their original contexts. The geographer Peirce Lewis (1979) has given a good idea of what this involves. He argues that the human-made landscapes are a valuable clue to culture, not least because they involve enormous investments of time, effort, and money; they look like they do for good reasons. If we look carefully we can usually discover those reasons, which may not always accord with the written and spoken claims of the builders. Lewis advises us that looking carefully requires us to treat almost everything as equal and connected; the homes of the rich and famous, trailer parks, skyscrapers, shopping malls, and lawn ornaments all have cultural significance. Such wide-ranging observation of ordinary things is astonishingly difficult for generations raised on textbooks and the expert opinions of others. Lewis is undaunted by this. He writes that one can "quite literally teach oneself how to see, and that is something most Americans have not done and should do." The alternation of looking, and reading, and thinking, he suggests, can raise questions we had not thought to ask, can reveal order in the landscape where we had seen only bedlam, and can otherwise yield remarkable results "that may be the road to sanity" (Lewis 1979:27).

What really matters about reading landscapes and developing a geographical sense of place is, I think, that these skills require us to learn to look critically for ourselves at the world and its landscapes. Our lives are awash with secondhand information and images, whether from textbooks or television, and to accept these without question is, in some measure, to give up our independence of thought. I cannot imagine a more rigorous test of the veracity of what we are told than that of our own observations of landscapes and places. For some, this may be an instinctive faculty, but for the great majority it is something that has to be learned and practiced. There is perhaps no better way to do this than through the type of geography that investigates particular settings, especially through fieldwork that is systematic, open-minded, and looks to see what might lie behind facades. A geographical sense of place teaches us how to interpret the complex grammar of environment, how to look for the elements of a place, their historical development and original contexts, and how to understand the interactions of land uses and social processes. With such a skill at our disposal there is little chance that we will succumb to environmental gullibility; and unlike King Lear we should never find ourselves mainly ignorant of what place this is.

A Poisoned Sense of Place

Much of what is positive in sense of place depends on a reasonable balance. When that balance is upset by an excess of placeless internationalism, the local identity of places is eroded. At the other extreme, when that balance is upset by an excess of local or national zeal, the result is a poisoned sense of place in that other places and peoples are treated with contempt. In other words, sense of place carries within itself a blindness and a tendency to become a platform for ethnic nationalist supremacy and xenophobia. This tendency was apparent, for example, in the place cocoons that Europeans took to protect themselves from the unacceptable local contexts of their colonies. Bits of Britain were reproduced in India, and a Spanish way of life was exported to Latin America. The tendency was also manifested in Nazi Germany, where an obsessive love of national landscape and culture led to the brutal attempt to purify the homeland by removing whatever and whomever did not belong.

In the past quarter century, instances of a poisoned sense of place have become pronounced. Place attachment in the guise of ethnic nationalism has thrust itself forcefully and often violently into the foreground of international politics. This sort of divisive parochialism had seemed destined to disappear with the growth of international organizations and global communications. Frontiers would be taken down, as in the European Community, and cultural differences would still be celebrated. Instead, a deeply paradoxical world is emerging, one in which cultural diversity is disappearing under an onslaught of global commercialism, even as old political affiliations rooted in history and place leap to the surface and are fiercely defended against ethnic outsiders. I have the impression from international television news that numerous areas in the world are filled with young men smoking American cigarettes and waving Russian rifles who want only to annihilate their cultural neighbors. In geopolitics it seems as though things are simultaneously coming together at a global scale and falling apart locally.

Place in ethnic nationalism is synonymous with the culture into which one has been born and therefore synonymous with the territory of ethnic symbols and associations where one belongs. In 1993 the political philosopher Michael Ignatieff visited several parts of the world where desperate searches for political identity were underway, most of them consumed with violence. He found a close connection between the revival of parochialism and the collapse of civil order. He asks rhetorically, "If violence is to be legitimated, it must be in the name of all that is best in a people, and what is better than their love of home?" As civil order disintegrates, safety and survival become paramount: "where you belong is where you are safe" and you are safest among your ethnic fellows (Ignatieff 1994:6). The stronger the belonging, the greater the hostility to outsiders. It is but a short step to the politics of ethnic cleansing and the forceful removal of others so that your place and your people are secure. A sense of place that stresses uniqueness to the virtual exclusion of a recognition of shared qualities is an ugly and violent thing. It is indeed a poisoned sense of place.

A Realometer—The Commonsense of Place

Michael Sorkin has edited a book about late-twentieth century cities titled (with remarkable relevance for Las Vegas that it nevertheless fails to discuss) *Variations on a Theme Park*. In his introduction Sorkin (1992:xi) speculates about the obsolescence of time and space, and the emergence of an entirely new kind of "ageographical city . . . a city without a place attached to it."

This news of the death of geography is greatly exaggerated. Sorkin has mistaken mere change for disappearance. The world is not shifting in its entirety to a web of invisible electronic impulses. Places did not vanish in the nineteenth century as the railroad and telegraph made remote places increasingly familiar, and, for at least three reasons, they will not vanish now. First, the boundaries between cultures have almost always been permeable to some degree, and the rapid global movements of peoples, ideas, and fashions that are occurring at the end of the twentieth century do not require that we abandon the ancient language of place, although they may require us to adapt it. Second, many premodern places remain, and a well-developed geographical sense of place can continue to enhance our appreciation of these by helping us to unravel the patterns of building and culture that give rise to their distinctive character. Third, tourism is now the largest industry in the world (if all the multipliers and air travel and so on are included), so we can be confident that distinctiveness in some form will be protected and created. Apart from anything else, a world without places and geographical variety would be boring and therefore not good for tourism.

The difficulty for Sorkin, as for many others who write of the geography of nowhere or the disappearance of places, is that their thinking and their language have failed to keep pace with changes in the world. The demanding challenge for an adaptable sense of place is to come to terms with how the world is changing *now,* and to do this it must understand the reconstituted geographies of postmodernity, sort out which geography is where, whose geography it is, whether it is largely an illusion, and whether any of what Strabo called virtue remains in it.

In his reflections about Walden Pond, Thoreau (1854: "What I lived for") imagined a Realometer, a gauge that would enable us to know "how deep a freshet of shams and appearances had gathered from time to time." He had in mind something that would enable us to know how deep we had to dig before we came to "a hard bottom and rocks in place, which we call reality." Such a device is much needed, but the difficulty now is that there seems to be no hard bottom and no clear reality. All geographies seem to be, to a greater or lesser degree, in flux, filled with illusions and fraudulence, exploitable and exploited for political or commercial ends. It is difficult to know where to begin to grapple with this turmoil, but one possible point might be at the extreme of place belonging. Michael Ignatieff (1994:186) thinks that ethnic nationalism embraces a plane of abstract fantasy and a plane of direct experience that are nevertheless somehow held apart. It is as though they are standing back to back, looking in different directions. The abstractions of fatherland, of purity of race, of superiority, are not allowed to confront the realities of shared experience in particular places. People, apparently, censor the testimony of their own experience so that they can believe in some abstract ideal. This is a powerful insight, one I suspect applies to much of post-modern life. What is needed to reduce this separation is a commonsense language of place that brings abstractions into correspondence with direct experience.

The *Oxford English Dictionary* is an excellent source of clarity in otherwise confused times. By common sense, it informs us in a quote taken from Leyland in 1651, "we usually and justly understand the faculty to discern one thing from another and the ordinary ability to prevent ourselves from being imposed on by gross contradictions, palpable inconsistencies and unmask'd imposture." Here is a possible basis for a realometer. There is nothing especially new in this; indeed, it is what geographers have taught in some fashion since Eratosthenes first struggled to organize the papyrus rolls in his library. A commonsense of place requires careful observation, critical reflection, and an awareness of interrelationships. I can think of no better, and perhaps no more challenging, task for geographers than to continue to teach these simple and important skills that they have long promoted.

Arguments for balance and reason have a long and excellent pedigree. This did not, however, prevent Clifford Geertz (1986:118), an ethnographer who is well attuned to the issue of diminishing cultural diversity, from describing this sort of concern for balance as a dribbling out to an ambiguous end. Given the geographical evidence of what happens when absolutes or extremes about place are promoted, I think his criticism is misguided. It nevertheless offers an important caution. Reason and balance in sense of place, as in many things, can slide into feeble ambiguities. To follow the middle road requires determination and an ability to resist the easy seductions of nationalism and place fabrication. I am convinced, however, that for the geographer exploring the transferred landscapes and reconstituted nationalisms of postmodernity, a common sense of place is an essential foundation for a balanced attitude of judgment that celebrates differences yet recognizes that there is much that different cultures can share without undermining their distinctiveness.

Note

I wish to thank Karen Reeds of Rutgers University Press for suggesting the term "poisoned sense of place," and Susan Hanson for her meticulous and thoughtful editorial comments.

References

Alexander, C. 1979. *The Timeless Way of Building*. New York: Oxford University Press.

Entrikin, N. 1991. *The Betweenness of Place: Towards a Geography of Modernity*. Baltimore: Johns Hopkins University Press.

Geertz, C. 1986. The Uses of Diversity. *Michigan Quarterly Review* 25(1):105–123.

———. 1988. *Works and Lives: The Anthropologist as Author*. Stanford: Stanford University Press.

Gropius, W. 1965. *The New Architecture and the Bauhaus*. Cambridge: MIT Press.

Harvey, D. 1989. *The Condition of Postmodernity*. Oxford: Basil Blackwell.

Hiss, T. 1990. *The Experience of Place*. New York: Alfred Knopf.

Ignatieff, M. 1984. *The Needs of Strangers*. London: Chatto.

———. 1994 *Blood and Belonging*. Toronto: Viking.

Jackson, J. B. 1970. Other-Directed Houses. In *Landscapes: Selected Writings of J. B. Jackson*, ed. E. H. Zube, 55–72. Amherst: University of Massachusetts Press.

Jacobs, J. 1961. *The Death and Life of Great American Cities*. New York: Vintage Books.

Lewis, P. F. 1979. Axioms for Reading the Landscape. In *The Interpretation of Ordinary Landscapes: Geographical Essays*, ed. D. W. Meinig, 11–32. New York: Oxford University Press.

McLuhan, M. 1964. *Understanding Media*. Toronto: Signet Books.

Meyerowitz, J. 1985. *No Sense of Place*. New York: Oxford University Press.

Relph, E. 1976. *Place and Placelessness*. London: Pion.

Ruskin, J. 1866. The Crown of Wild Olive. In *The Works of John Ruskin*, ed. E. T. Cook and A. Wedderburn, published 1903–1912. London: George Allen.

Sorkin, M., ed. 1992. *Variations on a Theme Park: The New American City and the End of Public Space*. New York: Noonday Press.

Strabo. 1917. *The Geography of Strabo*, in 8 vols., translated by H. L. Jones. London: William Heinemann.

Thoreau, H. D. 1854. *Walden*, 1960 edition. Boston: Houghton Mifflin.

Tuan, Yi-fu. 1974. *Topophilia: A Study of Environmental Perception, Attitudes and Values*. Englewood Cliffs: Prentice-Hall.

Venturi, R., D. Scott-Brown, S. Izenour. 1972. *Learning from Las Vegas*. Cambridge: MIT Press.

Vidal De La Blache, P. 1913. Des caractères distinctifs de la géographie. *Annales de géographie* 22:289–299.

Walter, E. V. 1988. *Placeways: A Theory of the Human Environment*. Chapel Hill: University of North Carolina Press.

Whitman, W. 1855. "Crossing Brooklyn Ferry," Leaves of Grass. In *Complete Poetry and Selected Prose*, 1959 edition. Boston: Houghton Mifflin.

About the Contributors

Elizabeth K. Burns is professor of geography and director of the Center for Advanced Transportation Systems Research at Arizona State University in Tempe. She received her Ph.D. in 1974 from the University of California, Berkeley. Integrating urban geography, planning, and transportation in her professional practice as a member of the American Institute of Certified Planners, she advised the City of Phoenix, Arizona, on its approved Urban Village comprehensive plan, served as a planning and zoning commissioner for Salt Lake City, Utah, and has conducted more than forty planning and project management studies for cities in Arizona, California, and Utah, as well as for the State of Arizona. She has examined the land use and transportation impacts of urban growth in southwestern cities in over fifty multidisciplinary scholarly publications, including articles in *Computers, Environment and Urban Systems, Transportation Research Record,* and *The Professional Geographer;* book chapters on urban growth, and the 1993 monograph, *Do Environmental Measures and Travel Reduction Programs Hurt Working Women?* She is currently investigating mobility barriers to inner-city employment in automobile-dependent cities.

Patricia Gober received a B.S. in geography from the University of Wisconsin-Whitewater in 1970, and M.A. and Ph.D. degrees from the Ohio State University in 1972 and 1975 respectively. She is currently professor of geography at Arizona State University, Tempe. Gober's areas of specialization are population and urban geography with more specific interests in residential mobility and migration, the geography of abortion, housing demography, spatial mismatch, and urban employment patterns. Recent publications include a 1993 Population Reference Bureau monograph, *Americans on the Move,* and numerous articles in the *Annals of the American Association of Geographers, Demography, Progress in Human Geography,* and *Growth and Change.* In 1994, Gober chaired a demographic analysis of employment conditions in geography that appeared as a three-article series in *The Professional Geographer.* In 1995, she gave the keynote address to the Association of American Geographer's Population Specialty Group dealing with new directions in population geography. Gober served as associate editor of the *Annals* between 1991 and 1993 and is currently on the editorial boards of *Urban Geography, Geographical Analysis,* and the *International Journal of Population Geography.* She is currently vice-president of the Association of American Geographers.

Anne Godlewska is associate professor of geography at Queen's University, Canada. She is passionate about maps and a lot of other things. Teaching is her life, but her publications include *The Napoleonic Survey of Egypt: A Masterpiece of Compilation* (available through the serials division of the University of Toronto Press); *Geography and Empire* (Blackwell, 1994), which was co-edited with Neil Smith, and numerous articles in English and French on the map and the evolution of spatial understanding; on the French scientific expeditions to Egypt and Algeria; on disciplinarity and geography in France (1790 to 1830); and on ideology as expressed in text, map, image, and art. She is currently working on three books: *Geography (Un)bound: When Description Fell to Theory, An Atlas of Napoleonic Cartography,* and *Napoleon and Europe.*

Michael F. Goodchild is professor of geography at the University of California, Santa Barbara, and director of the National Center for Geographic Information and Analysis. He received his B.A. degree in physics from Cambridge University in 1965 and his Ph.D. in geography from McMaster University in 1969. After nineteen years at the University of Western Ontario, including three years as chair, he moved to Santa Barbara in 1988. In 1990 he was given the Canadian Association of Geographers Award for scholarly distinction, and in 1996 the Association of American Geographers award for outstanding scholarship. He has twice won the Horwood Critique Prize of the Urban and Regional Information Systems Association. He was editor of *Geographical Analysis* between 1987 and 1990, and serves on the editorial boards of eight other journals and book series. His major publications include *Geographical Information Systems: Principles and Applications, Environmental Modeling with GIS,* and *Accuracy of Spatial Databases.*

Susan Hanson is professor of geography at Clark University. A graduate of Middlebury College, she earned a Ph.D. in geography at Northwestern after serving as a Peace Corps Volunteer in Kenya. She has written extensively on gender and local labor markets and on urban activity patterns. Her recent books include *Gender, Work, and Space* (with Geraldine Pratt) and *The Geography of Urban Transportation.* Professor Hanson is co-editor of *Economic Geography,* is a former editor of the *Annals of the Association of American Geographers* and *The Professional Geographer,* and currently serves on the editorial boards of five other journals. She is a past president of the Association of

American Geographers, a fellow of the American Association for the Advancement of Science, a former Guggenheim Fellow, and a recipient of the Honors Award of the Association of American Geographers. Professor Hanson currently serves on the board of directors of the Social Science Research Council, the Consortium of Social Science Associations, and the National Center for Geographic Information and Analysis. As co-chair of the AAG's Commission on College Geography, she also is leading a project to develop and disseminate active learning modules for introductory undergraduate courses.

Robert W. Kates is an independent scholar in Trenton, Maine, and university professor (emeritus) at Brown University. He is an executive editor of *Environment* magazine, co-chair of *Overcoming Hunger in the 1990s*, distinguished scientist at the George Perkins Marsh Institute at Clark University, faculty associate of the College of the Atlantic, and senior fellow of The H. John Heinz III Center for Science, Economics, and the Environment. Between 1986 and 1992 he directed the Alan Shawn Feinstein World Hunger Program at Brown University. Prior to 1986, Dr. Kates held various teaching and research posts at Clark University. He is a recipient of the 1991 National Medal of Science awarded by the president of the United States, the MacArthur Prize Fellowship, the Honors Award of the Association of American Geographers, and an honorary degree from Clark University. He is a member of the National Academy of Sciences, the American Academy of Arts and Sciences, and a fellow of the American Association for the Advancement of Science. In 1993 and 1994, he was the president of the Association of American Geographers. His recent books include the co-editorship or authorship of *Hunger in History: Food Shortage, Poverty, and Deprivation* (1989), *The Earth as Transformed by Human Action* (1990), *The Environment as Hazard*, (second edition 1993), and *Population Growth and Agricultural Change in Africa* (1993).

John R. Mather has been professor of geography at the University of Delaware since 1963. He served as chair of Delaware's Department of Geography from its formation in 1965 until 1989. Previous to that, he worked with C. W. Thornthwaite for sixteen years at the Laboratory of Climatology in southern New Jersey on the climatic water balance and its use in the solution of various applied climatic problems. On Thornthwaite's death in 1963, he directed the work of the laboratory for ten more years while also serving as chair of the Delaware Geography Department. A past president of the Association of American Geographers, Professor Mather is the author of two textbooks on climatology and the water budget and one on water resources. He has also coauthored *Global Change: Geographical Approaches* with Russian geographer, Galina V. Sdasyuk, and a biography of C. W. Thornthwaite with Canadian geographer, Marie Sanderson. He is the author of more than 100 journal articles and monographs on applied climatology, evapotranspiration, the climatic water budget, and water resources.

William B. Meyer is on the research faculty of the George Perkins Marsh Institute at Clark University. He is co-editor of *The Earth as Transformed by*

Human Action (1990) and *Changes in Land Use and Land Cover* (1994) and author of *Human Impact on the Earth* (1966). His research interests include the human dimensions of environmental change, urban geography, and the history of geographic and environmental thought. Meyer holds a bachelor's degree from Williams College and a doctorate in geography from Clark University.

Mark Monmonier is professor of geography in the Maxwell School of Citizenship and Public Affairs at Syracuse University. He received a B.A. in mathematics from The Johns Hopkins University in 1964 and a Ph.D. in geography from The Pennsylvania State University in 1969. He has served as editor of *The American Cartographer*, president of the American Cartographic Association, research geographer for the U.S. Geological Survey, and a consultant to the National Geographic Society and Microsoft Corporation. His awards include a Guggenheim Fellowship and a chancellor's citation for exceptional academic achievement from Syracuse University. He has published numerous articles on map design, automated map analysis, cartographic generalization, the history of cartography, statistical graphics, geographic demography, and mass communications. His books include *Maps, Distortion and Meaning, Computer-Assisted Cartography: Principles and Prospects, Technological Transition in Cartography, Map Appreciation, Maps with the News: The Development of American Journalistic Cartography, How to Lie with Maps, Mapping It Out, Expository Cartography for the Humanities and Social Sciences,* and *Drawing the Line: Tales of Maps and Cartocontroversy.* His current research projects include studies of cartography's role in controversial land-use decisions and a book on the evolution and significance of weather maps, broadly defined.

Edward Relph is professor of geography and chair of the division of Social Sciences, Scarborough College, University of Toronto. He is the author of *Place and Placelessness, Rational Landscapes,* and *Modern Urban Landscapes.* His *Toronto Guide* received a special award from the Association of American Geographers in 1990. Professor Relph's essays examine issues in geography, phenomenology, landscape architecture, urban design, and environmental philosophy; they have been excerpted in coffee table books and various magazines, including *Orion* and the *Utne Reader,* and translated into Russian, Japanese, French, Portuguese, Finnish, and several other languages. His Ph.D. is from the University of Toronto.

Edward J. Taaffe is currently professor of geography at Ohio State University, where he served as department chair for twelve years. Professor Taaffee is a past president of the Association of American Geographers and has served on the board of the Social Science Research Council. He was chairman of the National Academy of Sciences committee that produced the behavioral and social science survey report, *Geography,* and was a member of the National Academy committee that produced *The Science of Geography.* He has received the Honors Award of the Association of American Geographers, the Master Teacher Award from the National Council on Geographic Education, and was the first recipient of the Ullman Award in transportation geography from the

Association of American Geographers. Professor Taaffe's special fields are in transportation geography and geographic thought; his most recent book is *The Geography of Transportation*.

B. L. Turner, II, is the Milton P. and Alice C. Higgins Chair of Environment and Society at Clark University where he also directs the George Perkins Marsh Institute. In nine books and numerous articles, he has written on the relationships between nature and society, ranging from ancient Maya agriculture and environment in Mexico and Central America to contemporary agricultural change in the tropics and global land-use change. Professor Turner is a former Guggenheim Fellow, senior fellow of the Green Center for the Study of Science and Society, and fellow of the Center for Advanced Behavioral Studies. He is the recipient of Honors in research from the Association of American Geographers and is a member of the National Academy of Sciences. He has been deeply involved in the development of various agendas for the study of the human dimension of global environmental change. Turner was a member of the National Research Council's Committee on the Human Dimensions of Global Environmental Change and the Social Science Research Council's Committee for Research on Global Environmental Change, and he recently chaired the Core Project Planning Committee on Global Land-Use/Cover Change (LUCC) of the International Geosphere-Biosphere Programme and the Human Dimensions Program of the International Social Science Council. His current research projects include investigations of Amerindian agriculture in Middle America, agricultural change theory and the future of lands in the tropics, and the human causes of global land-use change.

Index

1
6-20

298
PBMT
E-70

66
CLB